DOWNRIVER

Orrin H. Ingram

and

The Empire Lumber Company

DOWNRIVER

Orrin H. Ingram
and
The Empire Lumber Company

CHARLES E. TWINING

THE STATE HISTORICAL SOCIETY OF WISCONSIN
MADISON • 1975

Manufactured in the United States of America by
Worzalla Publishing Company, Stevens Point, Wisconsin

Library of Congress Cataloging in Publication Data:
Twining, Charles E.
    Downriver.
1. Ingram, Orrin H., 1830–1918.
2. Empire Lumber Company.
3. Wisconsin—History.
I. Wisconsin. State Historical Society.
II. Title.
HD9757.W5T86   338.7′63′49820924   74–34404
ISBN 0–87020–149–2

PREFACE

WHATEVER ELSE history may be, it is always generalization. Although some subjects lend themselves more readily than others to investigation and discussion, no story is quite complete, no explanation entirely true. Without claiming a special challenge or complexity for the study of economic history in general or the lumber industry in particular, it is nevertheless true that the subject of lumbering has been paid scant and very tardy attention by American historians. Twenty years ago, in his article "Weyerhaeuser and the Chippewa Logging Industry," Paul W. Gates acutely observed that the *Dictionary of American Biography*, "among the many thousands of men whose careers it sketched, found space for only eight lumbermen . . . and among those eight is not found the one man above all who belonged there."[1]

This omission of Frederick Weyerhaeuser, and of many like him, was no mere oversight; the *DAB* was compiled far too carefully for that to have happened. Rather, it reflected a combination of factors: an apparent inability on the part of chroniclers to appreciate the importance of an activity so commonplace as making boards out of trees; the destruction of papers and business records in an industry more subject than most to floods and fires; and a peculiar reticence among certain of the most successful lumbermen. In time this led to the blurring, and virtually the disappearance, of their lives and works from the record of the American past.

Obviously it is the task of the historian to discern the men and events that truly counted, and to describe and analyze them in terms of the larger context. But this is more easily said than done. In some cases, important historical figures—financiers, political manipulators, Wall Street buccaneers—have deceived by conscious effort. In other cases, successful and influential men have simply managed to cast few public shadows. Their lack of visibility may occasionally

---

[1] Paul W. Gates, "Weyerhaeuser and the Chippewa Logging Industry," in *The John H. Hauberg Historical Essays*, ed. by O. Fritiof Ander (Rock Island, Illinois: Augustana Library Publications, 1954), 50.

v

have been the result of personal shyness; but more often these men
have held power so naturally, so confidently, that they required no
proof of it. We have come to admit such innate acceptance of
power and authority as typical among the scions of the New England
aristocracy; but somehow we find it difficult to imagine a Midwestern
lumberman wearing the same patrician robes with such ease and
grace. Yet why not? Frederick Weyerhaeuser apparently felt no
need to read of his own importance, and indeed went out of his way
to avoid publicity. "I am never interviewed," he told young Lincoln
Steffens; and although, as the journalist tells us, the lumber king did
finally consent to a lengthy discussion, it was with this proviso: "You
can't print it, of course."[2]

The neglect accorded lumbering and lumbermen by scholars reflecting on the American past was not deserved. Although prior to
the Civil War the industry was unusual in that it was highly decen-
tralized, there being more than 20,000 sawmilling and woodworking
establishments operational in the 1850's, taken as a whole it had
long been one of the nation's most important. In the mid-nineteenth
century, only the cotton-goods industry employed more people than
were employed in activities associated with lumbering, and the value
of forest products was exceeded only by the flour and meal industry.
In 1860, Wisconsin's lumber industry was (in the words of Frederick
Merk) "still in its infancy"; but already it ranked second to agriculture in importance.

The inattention paid the subject of lumbering is no more readily
explained by reason of any lack of romance or drama. Lumberjacks
were as rough, bold, and courageous as any other frontiersmen; the
lumbering entrepreneurs of the Lake States were as skilled, intelligent,
and daring as their big-city analogues back East. Furthermore, the
lumbering process itself is interesting, for it combines a variety of
activities which range from chopping down a tree in the wintry
isolation of a Wisconsin forest to pulling silt-covered boards from
the Mississippi in the muggy heat of a St. Louis summer.

Certainly, during the past half-century, subjects involving the American frontier have received their full measure of attention. Trappers
have trapped, miners have mined, farmers have farmed, and seemingly
each time that lonesome cowpoke climbed aboard Old Paint to chase
another longhorn, a new folk hero was born. Even the history of
public land policy, which some might maintain to be rather weari-
some, has been worked and reworked. But through it all, the lumbermen continued to chop and saw away in relative obscurity. Perhaps
Paul Bunyan was simply too big.

In recent years, academic historians have begun to focus a portion
of their attention on the forgotten chapters of lumbering; many of
their books and articles are acknowledged elsewhere in this volume.

[2] Lincoln Steffens, *The Autobiography of Lincoln Steffens* (New York: Harcourt,
Brace and Company, 1931), 365–367.

But lumbering has yet to receive its due in textbooks or in histories written for a more general audience. For example, the American Heritage *History of American Business & Industry,* a handsome, well-written, and supposedly comprehensive overview of the subject, contains no mention of Frederick Weyerhaeuser and only incidental reference to lumbering. A companion volume, *Great Stories of American Businessmen,* details episodes in the lives of inventors, real estate operators, and magnates of the oil, steel, railroad, automobile, and many other industries; but not a lumberman is to be found. Perhaps an even better example of this continuing neglect is provided by Daniel Boorstin's *The Americans,* the jacket of which promises "a new look at everything from Christmas to air conditioning. . . ." Professor Boorstin does indeed note all manner of subjects of varying importance from post-Civil War days to the near present, but again lumbering plays no part in the account.

The history of lumbering in America is intimately connected with the history of expansion, development, exploitation, and transportation. To a very great extent, lumbering *was* transportation. To have a real understanding of what was involved and how it was accomplished, one must appreciate the geographic relationship of pinery to prairie, and the technological aspects of the rivers and rails which connected them. (In 1953, William G. Rector dealt with this topic in his interesting study, *Log Transportation in the Lake States Lumber Industry, 1840–1918.*) Lumbering was also land and timber acquisition, manufacturing, distribution, and marketing within the frameworks of state and national laws. For the most part, these laws had not been written with forests or lumbermen in mind; but whatever their intent, they aided, or restricted, or gave direction to lumbering operations. (This aspect of lumbering history is ably discussed by James Willard Hurst in *Law and Economic Growth: The Legal History of the Lumber Industry in Wisconsin, 1836–1915.*) And finally, lumbering was men—individually and collectively, working together or competing fiercely, in relatively simple or in large and complicated organizations. (The human aspects of lumbering are illustrated by Ralph Hidy, Frank Hill, and Allan Nevins in *Timber and Men: The Weyerhaeuser Story.*)

What follows is the story of one of the more successful lumbermen, Orrin H. Ingram, and of his organization, the Empire Lumber Company of Eau Claire, Wisconsin. One of the earliest of Eau Claire's lumbermen, a young and ambitious Ingram arrived in the valley of the Chippewa in 1857, having already gained considerable experience in the forests of New York and Ontario. His Wisconsin career would outlast the pine, a seeming impossibility in the eyes of those who first viewed the extent of those timber resources. Because of his skill as a processor and business manager he soon built a modest empire. As a result, his responsibilities came to exceed involvement in his own company. Other lumbermen of the valley looked to him for leadership in their effort to deny their pine to the non-resident operators, led by Frederick Weyerhaeuser. Although Ingram and his associates were unsuccessful in this endeavor, their subsequent association with Weyerhaeuser proved extremely advantageous. But Ingram's success was not easily achieved, either in terms of personal sacrifice or in the demands he made upon others. Why he did it we can only surmise. That he did it was important to an expanding nation.

Realizing that most readers of this study will already possess a fairly complete understanding of the various operations which comprise the lumbering industry, I have not attempted to present a balanced, detailed description of each separate process. As a case in point, some may note, not unjustly, that the complications of timber acquisition have been largely neglected. I hope, however, that the particular strengths of the Ingram Papers, personal and corporate, will be apparent in the pages which follow. Specifically, I have emphasized, and discussed in considerable detail, the marketing aspects of the lumber business. The main title—*Downriver*—alludes to this.

*   *   *   *

With no intention of sharing the blame for any and all defects, I should like to acknowledge, with thanks, the assistance of a few of those who offered encouragement along the way. For whatever virtue this study may have, the credit is theirs.

Professor Andrew Clark of the University of Wisconsin introduced me, as a student of historical geography, to the subject of lumbering in the Lake States. Professor Vernon Carstensen, when he was at the University of Wisconsin, directed my early attempts at gathering information and expressing ideas with forbearance. My greatest debt is to Professor Morton Rothstein of the University of Wisconsin, who kindly but firmly put me back on the path when I tended to wander, tried his best to reduce my verbiage, and rescued me from pitfalls too numerous to mention.

I also want to thank Dr. Jerry Ham, Archivist of the State of Wisconsin, for his many favors, as well as Dr. Josephine Harper and her staff in the Manuscripts Library of the State Historical Society of Wisconsin. Professor Donald Woods, archivist and curator of the Area Research Center in the University of Wisconsin–Milwaukee, has tolerated my company for long periods with good spirit. Mrs. Doris Friedman, formerly librarian of the Eau Claire Public Library, has always been helpful in a most pleasant way. I must give special thanks to Professor Duane Fischer, historian and archivist in the University of Wisconsin–Eau Claire, for his generous assistance in all matters of research and also for providing such hospitable facilities and wholesome recreation along the banks of the Chippewa.

My wife Marilyn must be commended, not only for her many contributions but also for making it all worthwhile. In conclusion, I should like to thank Walter H. Ingram of Rice Lake, Wisconsin, grand-nephew of Orrin H. Ingram, for his thoughtful suggestions; and, most importantly, Charles H. Ingram of Tacoma, Washington, and Edmund Hayes of Portland, Oregon, for their patience, their generosity, and their contributions to my understanding of their grandfather.

CHARLES E. TWINING

Ashland, Wisconsin
August, 1974

CONTENTS

ILLUSTRATIONS follow page 148.

MAP of the upper Mississippi valley appears on page 33.

CHAPTER I

Early Experiences

FROM THE EARLIEST LANDFALLS, the forest was a dominant feature and a significant factor in the life of the North American settler. He needed lumber to build his houses and barns, churches and schools, and he needed wood for barrels and boxes, fireplaces, forges, and stoves. But since the Atlantic seaboard was heavily timbered from the coast inland to a distance unknown, the initial importance of the forest seemed primarily the result of its great extent. Indeed, the settlers came to look upon the trees in negative terms, more as an obstacle to their labors than a blessing, and they rushed to locate where they could take advantage of the occasional openings which occurred in the wilderness. These clearings, and not the trees, they regarded as having a special value. Largely from the character of the landscape, it was natural that the American frontiersmen came to view the forest as limitless and its wealth as inexhaustible.

Because the forest resource was available to all, and because the tools for processing trees into rough boards were available to most, the early settler was often his own lumberman. He cleared fields in order to raise his crops, and he used some but not all of the timber he removed. In a very real sense, the dense forests of North America came to delimit the forward edge of white civilization. Clearing the land had more than a little to do with progress, and in the process the men of England, "who grew oaks for profit," quickly adapted to the unfamiliar conditions of the New World.[1]

[1] Robert G. Albion, *Forests and Sea Power* (Cambridge: Harvard University Press, 1926), 232; Joseph J. Malone, *Pine Trees and Politics: The Naval Stores and Forest Policy in Colonial New England, 1691–1775* (Seattle: University of Washington Press, 1964). Charles F. Carroll has discussed the subject at length

1

Early in the pioneer experience, settlers began to specialize in various activities, including the production of lumber. The forests adjacent to lakes and streams became the scene of efforts to provide the lumber needed by neighboring communities. Sawmills were constructed at sites where the water ran swiftly enough to supply the power needed to drive the saws. Slowly and inefficiently, these early operators reduced the dark logs to piles of rough and uneven boards.

By the end of the eighteenth century, lumbering was an established business throughout the United States. Few communities of any size were without a sawmill and retail yard. These small concerns, along with blacksmiths, grist mills, and other stores and shops, often enjoyed nearly complete monopoly over a localized market. But as villages grew and communication between them improved, the old areas of trade began to expand and to merge. Although transportation was generally difficult, under certain conditions even bulky items such as logs and lumber moved between settlements, usually by means of natural waterways. With experience and increasing opportunities to be selective, buyers inevitably expressed a preference for the product of particular mills. The quality of their lumber or their comparatively lower prices, resulting from lower production costs, enabled some firms to enjoy an increase in demand. Accordingly, the managers of these relatively successful mills were able to gain the advantage in the changing situation. During this period the old-fashioned sawmill operator, accustomed to devoting but a portion of his working day to the production of boards and shingles, became a lumberman in a more modern sense. Concerned

---

in his recent book, *The Timber Economy of Puritan New England* (Providence: Brown University Press, 1973), the final paragraph of which reads: "The struggle against great odds is a recurring theme in early American history. And New Englanders, searching for traditions in a country still very young, were not merely sermonizing when they remembered those who once stood helpless and dismayed before the forest world and yet dared to enter in, change old customs and ways, and transform the wilderness." But Betty Flanders Thomson sees the obstacle as somewhat less imposing. In her article, "The Woods Around Us," *American Heritage*, Volume IX (August, 1958) 50–53, she observed that there were extensive open areas in the coastal forests at the time of initial European settlement. See also Lillian M. Willson, *Forest Conservation in Colonial Times* (St. Paul: Minnesota Historical Society, 1948), 1–3. For a discussion of the "Legend of Inexhaustibility," see Jenks Cameron, *The Development of Governmental Forest Control in the United States* (Baltimore: The Johns Hopkins Press, 1928), 1–9.

with improving the efficiency of his milling facility and with increasing its capacity, he began to resemble the entrepreneur of a familiar type.

Slowly, though at a steadily advancing rate, the lumberman exhausted his nearby source of logs. In order to continue operations, he was compelled either to move to another mill site or to establish logging camps in a more remote timbered region. But the forest could be of use only if streams provided the means by which to drive the logs to the sawmill. Along the coast and the banks of the coastal rivers, villages grew into towns and towns into cities, and lumbermen worked in new forests farther and farther removed from their expanding markets. By the third decade of the nineteenth century, inroads had been made into the timber resources of Maine, New Brunswick, New Hampshire, and New York—inroads of sufficient extent to force these lumbermen to look towards other sources of supply.

The pine regions of the upper Great Lakes were still too distant to be of any important commercial use, and some producers of the period moved into the more accessible forests of Pennsylvania while others regarded the Ottawa Valley of Canada as potentially the most valuable source of logs and lumber. That the forests might be Canadian while the major markets were American seemed to matter not at all. Man-made impediments in the form of political boundaries, land policies, freight-rate differentials, tariffs, and the like would be of relatively minor consequence in the character and progress of the lumbering frontier. Politics would not significantly alter the more fundamental factors of geography and economics, of timber resources and the demand for lumber.

The changes which occurred in the lumber industry during the first half of the nineteenth century were very small when compared with the changes which would take place in the fifty years to follow. If improvements in sawmilling machinery and lumbering methods had progressed at a leisurely pace during the earlier period, production nevertheless kept abreast the slowly increasing demand. When the farming frontier moved on to the treeless prairies of the trans-Mississippi West, however, the days of leisurely development abruptly ended. Rapid change became the natural order of things. Sawmill operators in the Lake States and the upper Mississippi Valley constructed new plants, designed new machinery, and adopted new techniques in their efforts to meet the seemingly insatiable demand for lumber. In the short space of five decades, from 1850 to 1900, the prairies and the plains would be settled and, not incidentally, the great pine forests of Michigan, Wisconsin, and Minnesota would be no more.

Lumbering was no longer of much importance in the vicinity of Southwick, Massachusetts, a small town near the Connecticut line a short distance west of the Connecticut River, when Orrin Henry Ingram was born there on May 12, 1830. Not many miles to the west, however, in the Lake George region of New York, Glens Falls was already recognized as a major sawmilling center. Orrin Ingram would soon catch up to the lumbering frontier and to the activity which would be his life's work.

In the early years of the century, grandfather David Ingram had left his home in Leeds, England, to settle in the rough and poor country of western Massachusetts. But the progress and prosperity he had dreamed about back in the Yorkshire hills came slowly and in small amounts in America. Although Orrin's father, also David, and his mother, Fannie Granger, had both been born in the Southwick area, neither felt bound there by any strong attachments. Instead, as their parents had done, they looked elsewhere in the hope of new and larger opportunities. Orrin was the fourth of nine children, six of whom reached maturity; and in providing for this growing family David Ingram had not done well in Southwick.

In the middle 1830's, when Orrin Ingram was still a small boy, the family moved to what they trusted would prove to be a better farm, located just east of Saratoga Springs, New York. This scenic country around Lake George was the childhood home that Ingram would later recall with fondness. But like so many regions of great natural beauty, much of the land seemed better adapted to forests than to farms. At best, father David's success on the new place was minimal. The older Ingram children early became accustomed to working for neighbors, contributing as best they could towards their own support. Many of their youthful memories were associated with long days of labor and lonely nights away from home.

When David Ingram died in 1841 at the age of thirty-seven, he left his wife the responsibility of a large family and little else.

Fannie Ingram had no choice but that the older children be "bound out," so eleven-year-old Orrin was sent to work and live on a farm near Glens Falls. Although the original agreement was that he would remain there until his twenty-first birthday, he stayed only two years. A short time after being widowed, his mother remarried and moved to a farm near Bolton, on the west shore of Lake George. Even before getting settled in her new home, she began to make plans and arrangements which would bring her scattered family back closer together. She soon succeeded in finding a neighboring family willing to board young Orrin in exchange for help with the chores, and, with the consent of the Glens Falls farmer, the change was made. Years later Ingram recalled that he had moved from his first position because of his mother's anxiety to have him near her. The thirteen-year-old must have been equally anxious for the reunion, and, as an additional benefit, his new employers lived close to a schoolhouse. Up to that time education in any formal sense had not been possible.

He had hardly begun to get acquainted in his new situation, however, before arrangements were completed for yet another move. An uncle, Nathan Goodman, related to Orrin by his mother's recent remarriage, expressed an interest in hiring young Ingram to work on his farm, about two miles distant from Bolton. Even though the Goodmans were complete strangers to the boy, they were nevertheless family, and Orrin had no choice but to pack his bag again. No matter how understanding and considerate the various families may have been, these early experiences were not the sort which encouraged a sense of security. Ingram doubtless began to appreciate many things at a relatively early age. One thought which must have occurred to him more than once was how very different his life would have been had his father enjoyed even a small measure of financial success.

Ingram would later remember that Uncle Nathan had a nice farm, a pleasant wife, and a considerable amount of work to be done. The Goodmans also had an adopted son, but this lad seemed rather more interested in the family library than in any strenuous outdoor activities. It required only a few days for Orrin to realize that he would not have much help with the chores. Although reluctant to charge his cousin with downright laziness, an older Ingram did suggest that "he was willing to let me do all the work while he would stay around the house and read."[2] But if Ingram the boy was anything like Ingram the man, he found considerable satisfaction in the performance of physical labor, and, he later recalled with obvious pride that summer days on the Goodman farm consisted of "man's work," even for a fifteen-year-old boy. Winter snows and school brought some changes to the daily routine, but the chores remained many and the hours long. He was often up at four o'clock in the morning and sawed by lantern light in the woodshed until called to breakfast. After the meal and morning prayers, he helped his uncle with the milking, fed the horses, and then got himself ready to start for the schoolhouse, a mile and a half away.

Without complaining specifically about the hard work and the long days, like many farm boys before and since Ingram considered most of the jobs "monotonous." Only the occasional opportunities to work with the few pieces of machinery that had gears and moving parts provided some relief from the boredom. It was the tools and machinery and the power of steam in engines that he found fascinating, interests not unusual for a boy in the 1840's. But on a farm one soon exhausted the lessons to be learned from such objects. Young Ingram also overheard discussions of Oregon and California and other places far away and full of promise, and it became an increasingly difficult matter to be satisfied with school work and chores.

Although the understanding with Uncle Nathan had also specified his twenty-first birthday as the date for terminating the arrangement, Ingram could not wait so long a time. His mechanical bent led him to correspond with another uncle, a brother of his mother who still lived in Southwick. This uncle claimed personal acquaintance with the commander at the armory in nearby Springfield, and it was in the hope of securing a position there that eighteen-year-old Orrin Ingram left Goodman's Corners.

[2] Orrin Henry Ingram, *Autobiography* (Eau Claire, Wisconsin: privately printed, 1912), 6. This *Autobiography* certainly is not complete, and ought to be used with an awareness of the spirit in which it was written. Ingram indicates that he intended it for family purposes alone. In any case, it is written in the style of one who considered himself a self-made successful businessman and does not mind calling attention to his humble beginnings and the agonies of each upward step. As is probably the case with most who begin such an autobiographical effort, Ingram soon tired of the task. It is important, however, because it is the only source for many of the early experiences.

But the plans did not work out. The Springfield Armory had no vacancies, and a long waiting list of applicants; the same was true at the locomotive works there. Uncle and nephew returned to Southwick, "feeling somewhat discouraged." Instead of learning a trade which involved new and exciting machinery and tools, Orrin was forced to accept a position as a clerk in a Southwick hotel, slightly augmenting his small salary by ringing the bell in the Congregational Church three times a day.

If chores on the farm had been "monotonous," work at the hotel desk was even more so, and in six weeks Ingram decided that he had had enough of Southwick. The best he could say for the experience was that for a brief period it had kept him "employed and out of mischief." Already he felt time slipping away, and was anxious to find a position in keeping with his interests and having some promise of advancement. As yet he had not given the lumber business any serious consideration although prior to leaving the Goodman farm he had been offered a job by Captain Harris, an old Lake George steamboat man, and Franklin H. Bronson, who together owned and managed the Harris & Bronson Lumber Company. From the vantage of his Southwick hotel desk, that offer now appeared more attractive than had earlier been the case.

Thus it was that Ingram left Southwick and returned to Lake George, from where he proceeded some twelve miles west of Ticonderoga to Lake Pharoah, the location of the Harris & Bronson sawmill. His first job in the mill, at a salary of thirteen dollars a month, was to operate a machine called an edger which squared the corners and sides of the boards. During the winter he cut logs in the woods for twelve dollars a month. Bronson took more than an ordinary interest in his new employees, no doubt in part because Ingram was on very friendly terms with Mrs. Bronson's sister, Cornelia Pierce. Cornelia, next-to-the-youngest daughter of Captain Pliny Pierce, had met Orrin when he was working on the Goodman farm and they had become more than casual acquaintances. But if Franklin Bronson recognized the possibility that here was a likely future brother-in-law, he was far more impressed by the fact that the young man so quickly demonstrated a thorough understanding of the sawmilling operation. Bronson handled the repairs, and in the evening after sawing was completed, he welcomed Ingram's company as he toured the mill, checking to insure that each piece of machinery was ready to resume operation in the morning. As they walked along, Bronson explained while his new employee listened and learned and thought of ways the work might be accomplished with greater speed, efficiency, and safety. It was not long—a matter of months —before Ingram was placed in charge whenever Bronson was absent. In addition to the experience gained in the mill, Ingram also assumed the management of winter logging operations for the firm. All in all, he received unusual opportunities to learn the business, and how to manage machinery and people to best advantage.[3]

Soon others in the area began to hear of Harris & Bronson's bright young man. Among them were a retired Presbyterian minister from Schenectady and a successful businessman and city treasurer from Kingston, Ontario. Together they had organized the Fox & Anglin Lumber Company in 1850 and shortly thereafter began the construction of a sawmill on the Rideau River, a few miles north of Kingston. Reverend Fox, whose money was the result of his marriage not his ministry, concluded that Orrin Ingram might be just the man to make a success of their new

[3] There are many sources for the biographical details of Orrin Ingram's family and life, including the Eau Claire newspapers published immediately following his death on October 16, 1918, as well as other Wisconsin papers such as the *Milwaukee Free Press* of October 20, 1918. A brief biographical sketch appears in the *Orrin Henry Ingram Funeral Memorial Book*, a copy of which is available in the library of the State Historical Society of Wisconsin in Madison. Also available at the SHSW library are the Marshall Cousins Scrapbooks, which contain clippings concerning nineteenth and early twentieth-century residents of Eau Claire and vicinity. References to Orrin Ingram are in Volume I of this collection. The various biographical collections of the period provide nearly identical information. See, for example, the Wisconsin Volume of the *Columbian Biographical Dictionary and Portrait Gallery of the Representative Men of the United States* (Chicago: Lewis Publishing Co., 1895); *American Lumbermen: The Personal History and Public and Business Achievements of One Hundred Eminent Lumbermen of the United States* (Chicago: American Lumberman, 1905), and William F. Bailey, editor, *History of Eau Claire County Wisconsin* (Chicago: G. F. Cooper & Company, 1914), 740–745. As indicated above, Orrin Ingram's *Autobiography*, written following the death of his wife in 1911, is the most important and complete account of his early life. See also Orrin Ingram to Platt B. Walker, Jr., March 9, 1900, in the Orrin H. Ingram Papers, State Historical Society of Wisconsin, Manuscripts Library, which are presently housed at the Eau Claire, Wisconsin, Area Research Center facility of the SHSW in the Library of the University of Wisconsin-Eau Claire. Unless otherwise indicated, all letters cited in this book are in the Ingram Papers at the State Historical Society of Wisconsin.

mill. His subsequent offer of an annual salary of $1000 plus board for management of the Rideau River sawmill was one that neither Harris nor Bronson cared to match, and one that the ambitious Ingram could not refuse.

The large and modern mill, designed to saw 150,000 feet of lumber a day, was not yet completed when Ingram arrived to assume his new responsibilities. His immediate attention, however, was directed farther upstream, where the crews were preparing to raft the winter harvest of logs to mills along the lower Rideau. Fox & Anglin had contracted for its supply of logs with an operator by the name of McDonald. In what proved to be a serious mistake, Anglin completed payment on the contract before receipt of the logs. McDonald subsequently departed in some haste without paying either his crew or his bills. Loggers and log drivers have traditionally tended to be less than timid in their dealings with others, even when they are in reasonably good spirits. McDonald's crews were in less than good spirits when they confronted young Orrin Ingram, demanding their pay.

Fox & Anglin's new manager doubtless matured in a hurry as he attempted to negotiate with "Bold McGinnis from County Tyrone," leader of the loggers who threatened to drive the rafts downstream, past the mill, selling to whomever would purchase for cash. The issue was clear enough: the mill could not operate without the logs. Ingram may have been somewhat uneasy facing such circumstances, but he appeared confident and firm; and with the assistance of local law enforcement officers, reinforced by a company of regulars, the Fox & Anglin logs ended up in the mill booms while "Bold McGinnis" and friends ended up with nothing to show for a winter's work. There were as yet no lien laws applying to logging operations in Canada; thus, the loggers had no legal recourse in such situations, which occasionally left them solely dependent upon the humanitarianism of the log owners. In this instance Reverend Fox and his partner chose to ignore their better instincts for the sake of business, a decision which Ingram apparently accepted with difficulty. Years later he recalled his misgivings, saying: "Although the Company was not at all liable, my way of treating such a matter would have been to do a little something for the men."[4]

---

[4] Ingram, *Autobiography*, 16.

Ingram worried less about business ethics at Brewer's mills, as the site of the Fox & Anglin operation came to be called, than he did about the climate, which seemed to encourage attacks of malarial fever or ague. Standing a husky six feet tall and possessing more than his share of strength and stamina, Ingram was seldom ill. But at this time he was becoming increasingly anxious about the amount of quinine which he had to consume in order to control the effects of the fever. This was a primary consideration in his decision to terminate his service with Fox & Anglin after a year and a half of superintending the Rideau River sawmill.

In the course of his employment there, a lumberman from Belleville, Ontario, about fifty miles west of Kingston, had asked him whether he might be interested in planning and constructing a sawmill for him. Ingram now decided to accept that proposition, because to design and build a mill was no ordinary opportunity for one so young. Even before this contract was completed, he had accepted offers to supervise the construction of two additional steam-powered gang mills in the vicinity of Belleville.

Had he been so inclined, Orrin Ingram might have looked back on a good many accomplishments in his first twenty-one years, but it seems unlikely that he took the time for such reflection. He was still a young man in a hurry. Despite his record of achievements, however, his life had not been all work, and before starting on the plans for the two new mills at Belleville, Ingram returned to Lake George to complete a bit of unfinished business there. What early understandings he and Cornelia Pierce may have reached we do not know, but it was certainly she that Ingram thought about as he strolled the shores of the Bay of Quinte during the summer evenings of 1851. They were married on December 11 of that year. The honeymoon was brief, Ingram hurrying back to begin work on the Belleville mills in order that they would be in operation the next spring or early summer at the latest. Cornelia stayed behind at Glens Falls with her sister, Mrs. Franklin H. Bronson, wife of the lumberman who had introduced Ingram to the business. Little time passed, however, before the newlyweds decided that they would prefer the inconveniences of Belleville to the loneliness of separation, and Ingram made the round trip again, this time returning to the north shore of Lake Ontario with his bride.

After completing construction of the mills, Ingram remained as the superintendent in one of them during the summer of 1852. He also accepted a contract to build a water-powered gang mill on the Moira River, nine miles from Belleville. Before the Moira River mill began sawing lumber in the spring of 1853, he received still another offer, this time from his former employers, Harris & Bronson. Well aware of Ingram's experiences and success since leaving their Lake Pharoah mill, they proposed that he build and manage a large mill for them at Bytown, soon to be known as Ottawa, Ontario. That Bronson and Ingram were now related was doubtless an important consideration for both, for in the nineteenth century, family ties commonly afforded such opportunities.

Harris and Bronson had not been the only New Yorkers to appreciate the potential advantages of producing Canadian lumber for the rapidly expanding American markets, but they were among the first to take hold of those opportunities. Captain Harris succeeded, for reasons not altogether clear, in negotiating an agreement with the mayor of Bytown, R. W. Scott, for the purchase of water power privileges, or hydraulic lots as they were called. Evidently the understanding simply specified that if Harris and his associates agreed to build a sawmill at Bytown, Scott would guarantee no adverse bidding, thereby assuring purchase of the water power at but one shilling more than the fifty-pound minimum price established by law. There were good reasons for the mayor's interest. In addition to capital the Americans brought skill and technical knowledge to the great forests of the Ottawa Valley, and Bytown soon became and long remained one of the continent's most important lumbering centers.[5]

Ingram likewise realized the potential of such an operation, and he readily agreed to return to the employ of Harris & Bronson without waiting upon the details of his new contract. His relationship with Bronson provided a sufficient guarantee of good faith, although in this instance such ties proved of little ultimate

[5] Arthur Reginald Lower, W. A. Carrothers and S. A. Saunders, *The North American Assault on the Canadian Forest* (Toronto: Ryerson Press, 1938), 112–115.

benefit. Bronson argued that Ingram ought to be given some financial interest in the Bytown mill. Partner Harris, however, obstinately refused. He might have been less opposed had Ingram then possessed his own funds to invest; but as matters stood, if he were allowed to take an interest, Harris and Bronson would be obliged to lend him the money to do so. Such an arrangement might be all right between members of a family, but Harris was not the young man's brother-in-law.

Ingram may have been unaware of these internal problems as he proceeded with plans and preparations, fully confident that the advantages soon to be enjoyed by Harris & Bronson would be second to none and that his own opportunities would be equally as promising. As for Cornelia, she was happy to be close to her sister again—and closer to civilization.

When completed, the Bytown mill was regarded as a standard for modernity and efficiency. Designed for a sawing capacity of 150,000 feet of lumber a day, Ingram later explained that the machinery and feeds had purposely been geared down with a view towards quality of production rather than quantity. By slight adjustments, the mill could have produced upwards of 300,000 feet per day.[6] Nearly all of the lumber sawed at the Bytown mill was loaded into canal boats for shipments down the Ottawa River to the St. Lawrence, to Lake Champlain on the Richelieu Canal, down to Whitehall at the southern extreme of the lake, and finally on to Troy by way of the Champlain and Northern canals. Thus, although the lumber was Canadian, the market was an old and familiar one for the firm of Harris & Bronson.

In recalling his own efforts in the new mill, Ingram admitted

---

[6] From the beginning to the end of the process—standing timber, sawlogs, mill capacity, and finally lumber—everything is measured in board feet. Board feet are to lumbermen what pounds are to cattlemen and bushels are to wheat farmers. One board foot is the equivalent of a piece of wood one foot square and one inch thick. Therefore, a twelve-foot-long 1 × 6 board would contain six board feet. (Purists will correctly observe that a 1 × 6 is actually only 25/32″ × 5 5/8″ after dressing, but that confuses unnecessarily.) The computation of board feet is largely an exercise in multiplication; that is, the number of pieces times the thickness in inches times the width in inches times the length in feet, the product of which is divided by twelve. To carry the examples further, for many years sawlogs in the Chippewa valley averaged between 190 and 200 board feet; and to construct a two-bedroom frame house today, it would probably require in the neighborhood of 10,000 board feet.

that the amount he accomplished seemed incredible, at least in retrospect:[7]

> I filed and hung every saw, including edging saws and trimmer saws; two saws for the first English or shore-gate, sixteen saws for the next, the slabber-part of the Yankee-gang, and twenty-two saws for the stock part of the Yankee-gang. I used a filing machine of my own invention, which enabled me to do the filing of so many saws as promptly as it had to be done. Then, at night, after the mill was stopped, I hung the saws I had filed during the day. I had a mill-wright who would go around with me to see that every key and everything else in the mill was properly cared for, to start the next morning. During that summer the only man besides the mill-wright I had to help me was one I took from the machine shop the month of August to help me about the filing. Other than that, I did all of that work myself, which would now be considered work for three men, at least. There were no machines then with emery wheels for sharpening saws. It was done with a file.

That the designer, builder, and superintendent of the mill personally spent so much time filing saws provides some indication as to the importance of that particular responsibility. Dull saws produced poor lumber.

The terms finally arranged with his employers stipulated that Ingram would receive a standard rate of seventy-five cents for each thousand feet of lumber, out of which he was to pay the expenses of running the mill, keeping the balance as his own salary. In a real sense, and for the first time, he was his own employer, and how much he profited depended upon how much and how efficiently he produced. At the end of that first season, when the accounts were totaled, Ingram found that he had earned an average of ten dollars a day for his work. Apparently this seemed excessive to Captain Harris, who offered to renew Ingram's contract for the coming season at the rate of fifty cents per thousand feet, a one-third reduction from their previous agreement. Ingram was surprised and disappointed at this proposed revision. He knew that he had done reasonably well for himself and extremely well for Harris & Bronson.

Under the circumstances, it was impossible for Ingram to remain in the employ of Haris & Bronson. The experience, how-

7 Ingram, *Autobiography*, 18–19.

ever, was not without its benefits. If he had not understood before, Ingram now had begun to appreciate the importance of recognizing the contributions of valued employees. Incentive had to be provided in order to satisfy the ambitious, and success in business depended upon the ambitions of all of those involved, not only the owners. Like Franklin Bronson in the 1850's, an older Ingram would one day be associated with business partners holding a very dim view of management personnel who expected to receive some material recognition for their contributions to a successful operation. To those future associates, as to Captain Harris, the investment of time and talent was clearly far from being equivalent to the investment of capital—and the division of the profits would inevitably reflect this conviction.

Again Ingram did not lack alternate employment opportunities. He had recently been approached by Gilmour & Company, then the largest producer of lumber in the world. In the course of their initial meeting, Allen Gilmour had merely indicated to Harris & Bronson's Bytown manager that if ever he was interested in considering a change, Gilmour & Company would welcome the opportunity to bid for his services. The Canadian headquarters of Gilmour & Company were also at Bytown, and, following his disagreement with Captain Harris, Ingram hurried to inform Gilmour that he was now "at liberty to engage with him." They lost no time in concluding a preliminary understanding.

These developments did not please Mrs. Ingram. As Orrin remembered the conversation with his wife when he told her of his decision to leave Harris & Bronson, he recalled that she was both shocked and saddened. Cornelia had hoped that he would remain associated with Bronson so that she could be close to her sister. "But," Ingram added with appropriate authority, "I was decided in my plan, and she, as she always has done, acquiesced."[8] Her objections were doubtless considerably lessened when they learned that if Orrin accepted the position with Gilmour & Company, they would probably continue to live in Bytown.

Before giving any final answer to Allen Gilmour, Ingram took the opportunity to look the situation over in detail. In the case of Gilmour & Company, this required considerable time and travel. The largest of the facilities was located on the Gatineau

---

[8] *Ibid.,* 20.

River, about nine miles north of Bytown, across the Ottawa River in Quebec. There the firm had sawmills situated at either end of a large dam, with a combined daily production capacity of 500,000 feet of lumber. It was proposed that Ingram assume direct management of the Gatineau mills and general management of the other Gilmour & Company mills in Canada at an annual salary of $4000, with the additional inducement that the firm would provide a house and horses. Gilmour obviously believed that this young man's talents would prove very important to the company and he was willing to pay a high price in return.

The holdings of Gilmour & Company were scattered throughout upper and lower Canada. Some two hundred miles from Bytown and west of Belleville there was a mill at Trenton on the River Trent, and in later years operations in that area north of Lake Ontario followed the lumbering frontier to Peterborough, Haliburton, and points farther north and west. In the Ottawa Valley, in addition to the mills on the Gatineau, Gilmour & Company had a mill at Buckingham on the Riviere du Lievre, one at Thurso on the Blanche River, smaller mills on the North Nation and South Nation rivers, and shipyard mills farther down the St. Lawrence at Wolf Cove and Indian Cove, Quebec. In addition to the lumber mills of Ontario and Quebec, they shipped large quantities of timber from New Brunswick, where the Gilmours were in partnership with John Rankin, that branch of the organization known as Gilmour & Rankin.

The interest which Allen Gilmour evidenced in Ingram resulted primarily from the latter's demonstrated talent for mill design and construction. Although he assumed many other responsibilities within the large organization, the new manager's special contribution involved his efforts at remodeling and improving the two mills on the Gatineau, as well as those on the Trent, the Blanche, and the North Nation, and the shipyard mill at Wolf Cove.[9]

In some respects, the experience with Gilmour & Company

---

[9] *Columbian Biographical Dictionary . . . of Representative Men of the United States: Wisconsin Volume,* 83. For a discussion of Gilmour & Company, its history, its importance and many involvements, including a general discussion of the nature of the trans-Atlantic lumber trade, see Arthur Reginald Lower, *Great Britain's Woodyard: British America and the Timber Trade, 1763–1867* (Montreal and London: McGill-Queen's University Press, 1973), 146–47.

was an entirely new one for Ingram. The Gilmours were neither native Canadian nor American. They were Scotch and English, and the basic orientation of Gilmour & Company was Old Country. At the other end of the line—in London, Liverpool, and Glasgow—Pollock & Gilmour was a famous and respected firm name. They transported Canadian lumber down and out of the St. Lawrence and across the Atlantic in their own fleet of ships. Numbering in the hundreds, each vessel flew a pennant bearing the initials of Pollock & Gilmour (which the sailors predictably charged stood for the "poor grub" prepared in the galleys).

When Ingram joined Gilmour & Company in the middle 1850's, little attention had as yet been given to either the local Canadian or the American markets. Indeed, the company mills produced few boards, sawing almost exclusively deals and square timbers, which would be resawed upon reaching English and Scottish ports. The square timbers were simply high-quality logs which had been squared to provide for more economical shipment. Deals were planks of three-inch thickness, or, as Ingram described them, "three inches plump," meaning a bit in excess of the exact measure.[10]

At that time the standard sawlog in the Ottawa Valley pineries was thirteen feet long. During the logging operations, the fallen pine trees were sawed into thirteen-foot sections whenever possible and, accordingly, the square timbers and the deals were most commonly that same length. The deals were sawed to vary in width at odd-inch intervals—seven, nine, eleven inches and larger—and were divided into three simple grades defining quality. No. 1 deals were nearly perfect, unblemished by either sap stain or knots; the middle or second grade permitted some discoloration and an occasional knot, although the lumber had to be entirely sound; and No. 3 deals allowed still more imperfections. But even the lumber sorted into the lowest grade was reasonably good since it was not practical to ship poor quality pine such long distances. Although prices differed considerably from year to year, in relative terms they remained much the same: No. 2 grades normally sold at two-thirds the price of No. 1; and No. 3 a third less, or one-third the price of the top-grade deals.

Naturally, in the process of fashioning square timbers and

---

[10] Ingram, *Autobiography*, 24.

three-inch deals out of round logs, a significant portion of the log was left behind, and the use made of these slabs and edgings had considerable effect on determining the amount of profit or sometimes of loss. What little lumber Gilmour & Company had been shipping to the Troy and Albany market consisted of 1″, 1¼″ and 1½″ boards which had been sawed from these left-over scraps and pieces. In other words, the Gilmours' involvement in the American markets was almost incidental, their orientation exactly the reverse of Harris & Bronson. This must have seemed a strange way of doing business to Ingram. He could not help but recognize the inefficiency inherent in producing deals and square timbers, especially at a time when there was such a demand for lumber in the adjacent and accessible New York markets. He was soon provided an opportunity to discuss this matter with his new employer.

Shortly after Ingram assumed the management of the Gatineau mills, lumber prices in foreign markets began to fall, especially on No. 2 and No. 3 deals. The decline became sufficiently serious that Ingram felt obliged to suggest to Allen Gilmour that they consider resawing their own low-grade deals for sale in New York instead of continuing to ship them to depressed Old Country markets. Gilmour agreed, permitting his new manager to go ahead with a design for a small gang saw capable of converting three and four deals at a time into boards. It was not long before the Gatineau mills were turning out large amounts of lumber bound for places nearer and better known to Ingram.

Lumber from the Gilmour & Company Gatineau mills followed the same route to Troy as did lumber produced at the Harris & Bronson mill at Bytown, and the two firms were soon locked in serious competition. As a result, although unintended, Ingram got a bit of revenge on his old employers. In any event, with the sudden introduction of large quantities of lumber from a new source, prices in the Albany-Troy market became depressed, lumber began to pile up; and Ingram and Gilmour soon discovered that in attempting to solve one problem they had created another. So much lumber was accumulating at Troy that Allen Gilmour had no choice but to send his Gatineau manager down to make arrangements for a yard at that location. This Ingram did, selecting an ideal site between and bordering the Mohawk and the Hudson rivers, thereby providing access to an extensive area. When sales were slow and time permitted, Ingram designed a system of tracks and cars with turn-table carriages to facilitate the handling of lumber in the yard. Although this was his first experience in the marketing phase of the business, the project proved a success. In its first full year of operation, the Troy yard of Gilmour & Company handled more than 30 million feet of lumber.

Orrin Ingram retained many pleasant memories of those busy days. The Gatineau mills became the showplace of the company, and managers from the other mills were brought to Bytown to see how Mr. Ingram handled his lumber, especially the product intended for sale in the American markets. In was apparent that Allen Gilmour held him in the highest regard, and not just in sawmilling matters. Indeed, they may have been as close as employer and employee could reasonably be. One experience which illustrated their relationship remained clear in Ingram's mind. It had occurred during one of Gilmour's visits to the Troy yard when he had hiked more than a mile in order to see the "lightning train" go by on the recently completed Hudson River road connecting Albany and New York City. Following this brief spectacle, Gilmour had hastened back to share the experience with Ingram, describing how he had stood on the quaking ground with his hands held tightly over his ears, awed by the size, power and speed of that great locomotive. "It seemed to demand an unconditional surrender of everything!" Gilmour explained, and Ingram had understood.[11] But there was an important limit to what they could share. Gilmour & Company was a family concern, and likely to remain so.

In meeting his more general responsibilities with the company, Ingram frequently traveled from mill to mill. It was necessary that he become familiar with each of the plants, always with a view towards increasing capacities and efficiency of operation at no loss of quality or safety. But as every successful manager knows, giving orders is seldom sufficient, and Ingram spent much of his time insuring that improvements had been completed in accordance with his specifications. In the course of these trips he naturally became acquainted with a great many of the employees of Gilmour & Company, one of whom, Donald Kennedy,

[11] *Ibid.,* 27.

soon became a valued friend and would later become a trusted business partner.

When they first met, Kennedy was foreman and millwright for the Blanche River sawmill at Thurso; and it was not long before Ingram went out of his way to enjoy a meal at the Kennedy table and spend an evening in pleasant conversation. About a year and a half older than Ingram, Donald Kennedy had been born at Harwell's Lock on the Rideau Canal, just north of Bytown. Although much less traveled than his frequent guest, Kennedy had the advantage of more formal education. But in spite of these differences in background, they shared a strong interest in machinery, as Ingram had appreciated during his initial inspection tour of the Blanche River plant.

In most respects, however, they were quite dissimilar, particularly as concerned their objectives and ambitions. Apparently Kennedy was content to continue to work for Gilmour & Company at the Thurso sawmill, taking pride in maintaining the machinery in the best possible condition. Ingram understood the satisfaction of organizing and managing a mechanical system which functioned smoothly and effectively, but to him such an accomplishment intrinsically was not enough. However well they ran, these machines still belonged to somebody else; somebody else made the important plans and divided the profits.

A solid and frugal type, Kennedy had managed to save some money out of his small salary, though in doing so he may have had no reason other than security in mind. Ingram was also beginning to accumulate a little capital, but for a more definite purpose—the ownership of his own lumber company. It seemed obvious to him that Kennedy would be an ideal partner in such an undertaking. Together they already possessed a great deal of knowledge in the sawmilling phase of the business, and Ingram had gained valuable experience in logging and marketing operations.

But their combined knowledge and experience could not make up for their lack of funds. The possibility of succeeding in the establishment of their own company in any settled region like the Ottawa Valley seemed very remote. There were, however, other forests to the west where pine was plentiful and inexpensive and where competing firms were fewer, smaller, and weaker. How much money would be required to insure a reasonable opportunity along the forward edge of the lumbering frontier Ingram did not know, but he now began to study the question seriously. It was already clear that were they to succeed in an independent operation, they would have to locate somewhere in the forests of the upper Great Lakes. Thus while Ingram studied and planned he also tried to broaden the horizons of his friend Kennedy, telling him stories of the big trees in the pineries of Michigan, Wisconsin, and Minnesota.

CHAPTER II

Western Fever

AS HIS FIRST YEAR with Gilmour & Company drew to a close, Ingram looked back with a few misgivings and looked ahead with some uncertainty. Aside from several thousands dollars he had been able to save, he seemed to have little to show for his hard work and many accomplishments. The tendency during such periods of personal assessment is to overlook the less tangible but extremely valuable considerations of experience gained and skills acquired. In truth, very few lumbermen twenty-five years of age had received opportunities equal to those enjoyed by Orrin Ingram. In the brief span of five years he had designed, constructed, remodeled, and superintended more sawmills than would most lumbermen in the course of an entire career. Furthermore, the mistakes he had made, which doubtless had been many, had been made with property and capital belonging to others.

By this time, however, Ingram considered his preparation and training quite adequate, and he was reluctant to give another year of his life over to Gilmour & Company. He recognized that as employers, the Gilmours were ideal. But for all their generosity and kindness, they were still his employers; and he was increasingly impatient to strike out on his own. Allen Gilmour appreciated the feelings of his Gatineau manager and, following the completion of some work remodeling the large steam mill at Trenton, he granted him two weeks' leave of absence so that Ingram might tour the pine sections of Michigan.

By the middle 1850's, there was already considerable activity in the Michigan forests, and when lumbermen spoke of unlimited opportunities and inexhaustible resources, they spoke of Michigan. But if Ingram was impressed by the standing pine and splendid sawmill sites, he was more impressed with some practical

21

matters which became apparent in the course of his discussion with Michigan operators. Of greatest importance, he learned that the amount of capital required to establish a sawmill which would have any fair chance of success, even in that frontier area, was considerably more than he had imagined. Welcomed back to Bytown by his wife and his employer, "satisfied that before attempting to go into the lumbering business in Michigan I ought to have more money than I had or knew where I could get it," Ingram readily accepted Gilmour's offer of a new contract with a raise of $2000, increasing his yearly salary to $6000.[1] Both Cornelia and Gilmour were pleased with the immediate results of Ingram's travels through the Michigan pineries, though it was clear to all that his plans had merely been postponed and not forgotten.

Once convinced that another year as an employee was the only reasonable arrangement, Ingram did not allow the disappointment to interfere with his work. His second year with Gilmour & Company was in many ways more productive than the first. Faced with less pressure and fewer crises than during the previous season, he had greater opportunity to think about the sawmilling processes in more general terms. Rather than spending so much of his time repairing old machinery, he was now able to devote longer periods to the design and construction of new and improved machines. The most important result of Ingram's continued employment with Gilmour & Company, at least as concerned the lumber industry, was his invention of the gang edger.

One of the earlier improvements in the production of lumber was the gang saw, a series of saws arranged parallel in a single frame. A large gang saw was capable of reducing a sizable log to a pile of boards at one pass. But no matter how the boards were sawed initially, it was usually necessary to improve their rectangularity, a function performed by a machine called an edger. It was natural for Ingram to focus his attention on that particular activity. In any system of production comprising separate but sequential steps, inefficient procedures soon become obvious. In the sawmill of the 1850's, the edging process was relatively inefficient. The speed of the over-all milling operation

---

[1] Ingram, *Autobiography*, 34–36.

suffered because of the slowness with which the edges of the rough boards could be squared.

For too long Ingram had watched boards and deals accumulate at the edger. No operator of that machine, regardless of his strength or skill, could keep up with the yield of the gang saws. He had to place the boards singly on a small table or carriage and direct them past the edging saw to square one side. Then he would pull the table back manually, turn the board over, and repeat the operation to square the other side. The solution which Ingram conceived to relieve the bottleneck was simple enough, employing the principal of the gang saw in the mechanism for edging. It seemed reasonable that by mounting a second edging saw on an adjustable arbor or shaft and connecting a pinion to the moving table, both edges could be squared simultaneously and automatically. A millwright named John R. Booth, later to become Canada's most successful lumberman, assisted Ingram in making a model of the gang edger so that he could better explain it to Allen Gilmour.

As usual, Gilmour was willing to accept his manager's recommendation. Following their conference, Ingram proceeded directly to the machine shop which handled the Gilmour & Company business and placed an order for five of the new machines, three of which were intended for his own Gatineau River mills. He later observed, with understandable pride, that his gang edger had "proved to be a grand success."[2]

In this instance, Ingram proved to be a better inventor than a businessman. For one reason or another, he neglected to apply for a patent on the new device until it was too late. Although he subsquently frustrated the attempts of others to claim the invention and to secure patent rights for themselves, Ingram personally profited not a penny from the rapid and widespread adoption of the gang edger, which was soon an indispensable part of every modern mill. Too generously, the authors of the many biographical sketches of Ingram's life attributed his failure to apply for a patent to some higher and unselfish motive.[3] In truth,

[2] *Ibid.*, 37.

[3] For example, in the biographical volume published by the *American Lumberman* in 1905, the practice supposedly followed by a member of the medical profession—that of sharing his discoveries for the good of all mankind—was contrasted with that of the venal businessman, traditionally intent on hoarding

if Ingram had one regret concerning his early years, it was his failure to follow the advice received and obtain a patent on the gang edger. He must have kicked himself a great many times when estimating what that lost opportunity might have been worth. Sometimes he placed the value forfeited at 15 million dollars and other times at twice that amount; but whatever the exact figure, at the turn of the century an older and wiser Ingram never doubted that the patent rights would have been "worth more to me than the best lumber establishment in this or any other country."[4]

During 1856, Ingram was far too busy with company business and personal plans to worry much about the possible results of neglecting to patent his invention. At every opportunity he talked of what he had seen and heard in Michigan, and he listened to others who had ideas and information concerning lumbering in the forests to the west. Occasionally he came across someone possessing plans similar to his own. Everyone in Ingram's situation was searching for a partner with capital, and it had to be something of a delicate matter attempting to discover whether your new acquaintance had a bank account to go along with his interest in the pineries of the upper Great Lakes. One such acquaintance who seemed more promising to Ingram than most was an older gentleman by the name of Alexander M. Dole. Dole was then in the employ of Hamilton Brothers, a lumbering concern with a mill at Bytown and headquarters at Hawkesbury, some distance down the Ottawa.

The two actually had little in common, save a mutual interest in going into business on their own, and for such an undertaking both looked toward the same distant forests. In 1856, however, the dreams they shared were sufficient to countervail their more obvious differences. Dole was considerably older than Ingram, but his experience in the lumber industry had not been nearly as extensive. Throughout Dole's long association with Hamilton Brothers, his responsibilities had largely been restricted to land procurement, tax questions, and logging operations. But if he lacked either knowledge or experience, he was not the sort to

---

for himself the profits which resulted from his good fortune, "unless, like Mr. Ingram, he thinks more of improving methods and benefiting the trade than he does of harvesting great results for himself alone."

[4] Ingram to Platt B. Walker, Jr., March 9, 1900.

admit it to Ingram or to anyone else, especially not to anyone younger. He operated on the premise that older men were inevitably wiser men.

Although Dole was clearly a bit of a character, Ingram was too anxious to get the project under way to be much concerned with seemingly minor personality quirks. Ingram had worked with a great many characters over the past few years, and for the most part successfully. Furthermore, little peculiarities could be overlooked or at least tolerated, especially if they were balanced by a particular strength or skill. In this connection, Ingram was aware of some deficiencies in his own background. He had, for example, little experience as yet in logging activities and almost none in land acquisition, subjects which had concerned Dole for many years.

Partners could be important for a variety of reasons, but at the moment, Ingram's paramount requirement was for additional money. Alexander Dole possessed considerable capital. Accordingly, whatever his shortcomings may have been, he looked like the ideal partner to Orrin Ingram. Plans proceeded rapidly, and when Dole left the Ottawa Valley on a fact-finding tour through the Minnesota Territory in the fall of 1856, he took with him the tacit understanding that should he find what seemed a promising situation, and should Ingram agree with his selection, they would make the effort together.

Dole traveled directly to St. Paul, and in that vicinity he had his first opportunity to view the stands of white pine in the upper Mississippi Valley. He was not disappointed by what he saw. Soon after arriving, however, he began to hear stories about a region in Wisconsin with bigger pine, better streams, and, as yet, an unlimited choice of mill sites and timber land. Responsible for most of the noise was Eau Claire's first town-boomer, a mechanic recently arrived from Madison, Adin Randall. At least in the case of Dole, Randall's advertising campaign worked, for the visitor became sufficiently curious to make the difficult journey to the frontier settlement on the Chippewa River.

Whether Alexander Dole saw the situation through his own eyes or through those of the enterprising Randall is open to question, but the old gentleman was soon convinced that Eau Claire had to be the best possible place to settle. Adin Randall just happened to have a property on the west bank of the river, complete with a small portable sawmill, which he was willing to sell. When Dole left Randall and Eau Claire for home, he had in his pocket an option on the mill and the site.

Ingram was predictably pleased with Dole's enthusiastic report, and he hastened to make plans to go to Wisconsin in order to survey matters for himself. In truth, the possibilities would have had to have been barren indeed to discourage him. Regardless of the accuracy of Dole's report, it was most unlikely that Ingram would again defer his dream. He was as anxious to buy and to begin as Randall was to sell, and before starting on the trip west Ingram gave Allen Gilmour three months' notice. Unless something quite unexpected occurred, he apparently assumed that once on the scene he would close the deal with Randall and immediately proceed with the construction of a modern sawmill.

In spite of his relationship with Dole and the recent developments which had resulted, Ingram had not forgotten Donald Kennedy. He had, in fact, figured quite prominently in Ingram's plans, although these had not as yet been entirely shared with him. In addition to a knowledge of the millwright's mechanical skills, he also knew that Kennedy was underpaid by Gilmour & Company and was therefore likely to give favorable consideration to a change, even one which necessitated a move far from home. But unlike Ingram, Kennedy was a victim of inertia, and it was only his confidence in his friend Ingram's judgment that moved him to become involved in the enterprise. Initially Ingram merely suggested that Kennedy accompany him to Eau Claire where they could look things over together, the entire trip requiring no more than three or four weeks. Kennedy finally agreed to this, on the condition that Ingram would find someone to assume the responsibilities at the Blanche River mill. A replacement was shortly found, and in mid-winter the two friends started for Eau Claire. Many years would pass before Donald Kennedy would return to the Ottawa Valley.

It was February when they arrived at their destination. Logging activities were in full swing, and Randall took his customers a short distance up the Eau Claire River to see some of the work going on in the woods. If Ingram was not an expert logger, he nonetheless appreciated the properties of a good sawlog, and he was impressed with the pine along the Eau Claire. When he shared this impression with the others, Randall responded by telling them that it was probably the poorest lot around, adding that if only they had more time he could show them some really big trees farther up the Chippewa. Ingram and Kennedy must have smiled at the possibility that these early Wisconsin operators were voluntarily logging the poorest stands first, reserving the choice areas for late-comers. But they saw enough to be convinced that Dole had made no mistake in his assessment of the valley.

The frontier settlement of Eau Claire was lacking in most refinements. There was as yet not even a church; there was a minister, however, and another was soon expected, and regular services were already being conducted in a little frame house. Such matters were important to the Ingrams. Pine without preaching would have been difficult for Orrin and intolerable for Cornelia; but with plenty of one and the promise of more of the other, Ingram believed that "things were looking rather promising."[5]

The partners wasted no time in deciding that Kennedy should remain in Eau Claire. Using the portable mill purchased from Randall and some logs available near the mouth of the Yellow River, Kennedy could begin sawing out the lumber and the timbers needed for construction of their permanent sawmill as soon as the weather would allow. In the meantime, Ingram would return to the Gatineau River mill and complete his final three months with Gilmour & Company.

While the opportunities for any detailed survey of the western pine regions had been limited by their own impatience and the difficulties of travel, it is doubtful that a more extensive examination would have resulted in a more favorable situation. They had located near the southern edge of an immense forest which was adjacent to a seemingly limitless prairie. The second of these two features, the prairie, was a relatively unfamiliar element in the American frontier experience. From colonial times, settlers had faced and fought the trees which impeded their efforts to make farms, build towns, and cut roads. But as the westward advance continued towards the Mississippi, the forest became less and less an obstacle. Larger and larger openings appeared with increasing regularity. And when the river barrier was crossed and settlement spread over the fertile prairies beyond, the infrequent groves of trees were quickly felled, no longer for pur-

---

[5] Ingram, *Autobiography*, 40.

poses of clearing the land but to provide for the needs of the settlers. The produce of the forest, ubiquitous along successive eastern frontiers, was almost totally lacking in the trans-Mississippi west.[6] Such a deficiency over so large an area gave added significance to the forest stretching across Michigan, Wisconsin, and Minnesota. Of equal importance, the streams of the upper Midwest flowed from the forest down towards the fertile farmland.[7]

The task of supplying this treeless region with cheap and abundant quantities of good wood was one which lumbermen assumed gradually. As with the other American frontiers, lumbering generally progressed in a westerly direction. Thus, in the upper Great Lakes forest, the pineries of Michigan absorbed the initial logging activities of any major consequence; and those of northern Minnesota, the last in the east-west line of advance, were the last to fall. As early as 1809, a sawmill was operating at De Pere, just south of Green Bay on the Fox River.[8] By 1830, lumbering operations were being carried on along the banks of most of Wisconsin's useful logging streams. At mid-century, however, the

---

[6] It seems unusual, at least in retrospect, that even in Wisconsin, a state which would lead the nation in lumber production before the end of the nineteenth century, this "lack of forests" proved a considerable inconvenience to early settlers. Not only did lead miners in the southwestern corner of the state live like "badgers" in caves, but the pine lumber used in constructing the first Wisconsin Capitol at Belmont had to be imported from Pennsylvania. An interesting piece of propaganda, aimed at counteracting the poor impression resulting from the lack of timber in southern Wisconsin, is found in William Rudolph Smith's *Observations on the Wisconsin Territory, Chiefly on that Part called the Wisconsin Land District* (Philadelphia: Carey & Hart, 1838.)

[7] It should be noted that these conditions—the pine resource, the prairie markets, and the inexpensive and relatively efficient method of connecting the two by means of natural waterways—were not the only factors which encouraged the rapid cutting of the Lake States forest. Additional reasons, obvious and often discussed, included timber trespass and the costs involved in attempting to protect pine stumpage from unauthorized cutting; the constant danger of destructive forest fires; and taxation rates on timber holdings, which were often intentionally unfair to absentee landowners. While acknowledging the importance of these and other factors, it should be understood that they were ancillary and not the basic causes. The trees would have been cut with dispatch even had there been no timber thieves, forest fires, or local taxes. For a discussion of the rapid exploitation of the Wisconsin pineries, see Charles E. Twining, "Plunder and Progress: The Lumbering Industry in Perspective," in the *Wisconsin Magazine of History*, 47 (Winter, 1963–1964), 116–124.

[8] Reuben Gold Thwaites, *Wisconsin: The Americanization of a French Settlement* (New York: Houghton Mifflin Company, 1908), 281.

timber that had been removed was but a trifling amount, and in 1860, "the invading loggers were yet thundering only at the outer gates of the vast forest solitudes of Wisconsin."[9]

When Wisconsin entered the Union in 1848, its forests covered an estimated 16,900,000 acres. The Wisconsin pine lands ranged generally north of a line from Green Bay to the mouth of the St. Croix, including nearly one-half the area of the state. Within that region, and among all of the tributaries of the upper Mississippi, the Chippewa valley comprised the largest and single most important lumbering district. Draining more than a third of the Wisconsin pinery, its original stand was estimated to have exceeded 46 billion board feet—16 billion feet more than the next largest district, the Wisconsin River valley.[10] Not only did the Chippewa basin contain more pine than was growing in the neighboring valleys, but the greater portion of that pine was unusually accessible to the lumbermen. Its lakes, rivers, and streams formed a natural network of waterways, providing excellent transportation for a bulky product. The valley of the Chippewa was, indeed, "a logger's paradise."[11]

With its principal tributaries, the Flambeau, Menomonie (Red Cedar), Eau Claire, Jump, Yellow, and Court Oreilles rivers, the Chippewa embraced within its broad basin more than 9500 square miles, which amounted to nearly one-sixth of the state. The low, rolling hills and drainage pattern of the valley, containing hundreds of small lakes and large expanses of swamp and marsh land, were not well suited to standard agriculture. Although occasional fertile pockets occur within the basin, the surface cover of most of the area consists of thin sand or sandy loam, soils for the most part favorable to coniferous and mixed forest growth.[12]

[9] Frederick Merk, *Economic History of Wisconsin During the Civil War Decade* (Madison: State Historical Society of Wisconsin, 1971 ed.), 60.

[10] John T. Curtis, *The Vegetation of Wisconsin* (Madison: University of Wisconsin Press, 1959), 171–172. Much earlier, Filibert Roth estimated that the original forests of the state included some 30 million acres containing at least 200 billion feet of saw timber. Filibert Roth, *Forestry Conditions of Northern Wisconsin* (Madison: Published by the State, 1898), 6–16.

[11] Frederick E. Weyerhaeuser, quoted by Ralph W. Hidy, Frank Ernest Hill, and Allan Nevins in *Timber and Men: The Weyerhaeuser Story* (New York: The MacMillan Company, 1963), 43.

[12] It is an obvious observation that had the soils of northern Wisconsin been better suited for agriculture, there would have been much less criticism of the

Romantic descriptions to the contrary, the pineries were not dense forests of stately white pine to the exclusion of other species. The best pine came from stands in which pine predominated, but was intermixed with hardwoods. Indeed, tracts covered exclusively with pine were seldom found, and the white pine rarely amounted to more than 60 per cent of the growing trees. The composition of the Great Lakes forest varied considerably from area to area, largely depending upon soil and moisture conditions. From federal land surveys, descriptive accounts of early travelers, and various other sources, it is possible to reconstruct, "with fair accuracy," the forest of northern Wisconsin as it existed prior to any large-scale lumbering activities:[13]

> The wet lands contained either conifer swamps, dominated by tamarack, black spruce, and white cedar, or hardwood swamps with black ash and yellow birch. The dry lands were dominated by pine, with jack and red pine on the lighter sands and the white pine on the sandy loams. The heavier soils were typically covered by mixed conifer-hardwoods, with white pine, hemlock, balsam fir, and white spruce as the conifers, and sugar maple, basswood, yellow birch, American elm, red oak, and ironwood as the deciduous species.

The white pine was not always the most common tree in the forest, but it was by far the most important. This pre-eminence resulted from its superior qualities. The wood is soft and light, yet strong. It can be easily worked by man and is adaptable to most wood-working purposes. It is also extremely buoyant, a quality of foremost consideration to the lumberman who floated his logs to mills and his lumber to market.[14] On streams large

lumberman's avarice by the more recent conservationists. Indeed, the loggers would have performed a valuable, if unintentional, service by clearing the land, thus permitting the easy establishment of farms. But despite early dreams and later propaganda, the farm has been unable to follow the forest into the majority of central and northern Wisconsin areas. See Vernon Carstensen, *Farms or Forests: Evolution of a State Land Policy for Northern Wisconsin, 1850–1932* (Madison: University of Wisconsin College of Agriculture, 1958); Arlan Helgeson, *Farms in the Cutover* (Madison: State Historical Society of Wisconsin, 1962); and Erling D. Solberg, *New Laws for New Forests* (Madison: University of Wisconsin Press, 1961), 22–41.

[13] Curtis, *Vegetation of Wisconsin*, 177. See also *A Century of Wisconsin Agriculture*, Wisconsin Crop and Livestock Reporting Service, Bulletin No. 290 (Madison: Published by the State, 1948), 6.

[14] William R. Raney, "Pine Lumbering in Wisconsin," in the *Wisconsin Magazine of History*, 19 (September, 1935), 71–72. Only cedar exceeds the eastern white

and small, wherever water ran in the desired direction and wherever spring freshets occurred naturally or could be created, the pine was moved; and no waters tributary to the Mississippi would carry more logs and lumber than those of the Chippewa River of Wisconsin.

In the myriad lakes and connecting swamps of northwestern Wisconsin, some of which are no more than twenty miles from the shore of Lake Superior, lay the sources of the Chippewa system. The elevation at the headwaters of the Flambeau—actually the main line of drainage in the upper basin—is more than 1600 feet above sea level. The river falls to 1050 feet at the confluence of the Flambeau and the Chippewa, to 806 feet at Chippewa Falls, and to 665 feet at the mouth, where the Chippewa joins the Mississippi.[15] Despite a difference of nearly 1000 feet from source to mouth, there are few waterfalls and cascades in the main streams. Above Chippewa Falls, frequent stretches of rough water or "boulder rapids" do occur. But in the lower river, although the gradient exceeds two feet per mile and the water runs swiftly in most seasons, rapids are a rarity.[16] This combination of strong but seldom turbulent current was to be of inestimable value to the lumbermen.

Although a great many dams would be constructed on streams throughout the valley, for the most part their function was one of conserving water to be released during the log drives, rather than the creation of a head of water for purposes of power. Considerably more important than water power to all but the earliest lumbermen in the valley was the availability of protected stretches of water where logs could be retained until required by the saw-

---

pine in those properties considered important for most construction purposes and many other woodworking uses. See Reginald D. Forbes,: *Forestry Handbook* (New York: Ronald Press Company, 1955), Table 13, "Approximate Comparison of Seven Properties Important in the Processing and Use of Wood." Air-dried white pine weighs on an average some twenty-six pounds per cubic foot. Cedar is the only wood commonly used that is lighter. It weighs twenty-two pounds p.c.f. By comparison, red pine weighs twenty-eight pounds; seasoned Douglas fir, thirty-four pounds; and white oak, forty-seven pounds p.c.f. *Timber Construction Manual* (Ottawa: Canadian Institute of Timber Construction, 1963 ed.), 266.

[15] William H. Herron, "Profile Surveys of Rivers in Wisconsin," in *United States Geological Survey*, Water Supply Paper 417 (Washington: Government Printing Office, 1917), 7–8.

[16] Lawrence Martin, *The Physical Geography of Wisconsin* (Madison: Published by the State, 1932 ed.), 201.

mills. On a stream as rapid as the Chippewa, however, such storage opportunities were uncommon in the natural state; and the attempts of man to improve the natural state in this respect succeeded only with difficulty and at large expense. Indeed, viewed in its simpler aspects, much of the history of the Chippewa valley can be reduced to a description of the efforts to provide bigger and safer areas for holding sawlogs. It was this concern that was fundamental to the long and bitter struggle between the sawmill operators of Eau Claire and Chippewa Falls.[17]

It proved to be an unequal contest from the beginning. Despite being downstream from Chippewa Falls, Eau Claire enjoyed far superior natural advantages. Long ago, the river had left behind an oxbow when the main stream suddenly straightened its course, forming Half Moon Lake. Both ends of the placid, two-mile-long lake were separated from the newer channel by slender necks of land. Claimed by a local editor to be "capable of holding a sufficient number of logs to supply half the United States," Half Moon Lake was ideally suited to the requirements of local lumbermen.[18] Its usefulness was assured early in 1857 when Adin Randall and Daniel Shaw completed excavation of a 1200-foot canal connecting the lake with the Chippewa River. That improvement bordered the property recently purchased by the new firm of Dole, Ingram & Kennedy.

When Ingram left Kennedy and Eau Claire to return to the Ottawa valley in February, 1857, it was understood that his stay with Gilmour & Company would be no longer than three months, as he had agreed upon with Allen Gilmour. Thus, if all went according to plan, Orrin and Cornelia Ingram would arrive on the banks of the Chippewa by the first of May. But everything did not go according to plan. A personal matter caused another brief postponement. The unexpected death of David Gilmour, a brother of Allen and the manager of the Quebec office of Gilmour & Company, required his brother Allen Gilmour to make the annual trip to England, which made the planned departure

[17] The most complete description by a contemporary of the Chippewa River and its tributaries as driving streams are the letters from Charles H. Henry to William H. Bartlett of March 3 and 5, and May 18, 1928. A minor Civil War hero, Captain Henry was a respected woodsman who worked for both Ingram and Frederick Weyerhaeuser. The letters are in the Bartlett Collection of the Eau Claire Public Library.

[18] *Eau Claire Telegraph*, May 25, 1857.

of Ingram especially difficult for Gilmour & Company. Lacking
alternatives, he asked Ingram to consider staying on at the Gatineau mills, at least until his own return from England. Ingram
could hardly leave under such circumstances, and he agreed to
remain with the firm during the period of Allen Gilmour's absence.

Even with the delay, the Ingrams reached Eau Claire on June
11. Although the season was ideal for taking up residence, there
was little else in the community to cause enthusiasm and Cornelia
doubtless began to fear that she had seen the last of civilization.
The muddy streets were filled with pigs and lined with saloons.
Still, the Congregationalists and the Presbyterians were now
both established and active, so all was not complete darkness. As
for Orrin, he was very satisfied with the progress Kennedy had
made on the new mill. It even began to seem possible that Dole,
Ingram & Kennedy might offer some lumber for sale that same
season.

All things considered, and allowing for the apparently inevitable fluctuations in nineteenth-century money matters, the three
partners could hardly have selected a better time and place in
which to begin their independent operation. Situated in the
midst of an abundant resource, adjacent and accessible to what
would soon become an insatiable market, their success in the
lumber business appeared more than likely. At the time, of
course, it was not all that certain. Lumber had to be made at low
expense and sold at low prices, in competition with many other
producers who enjoyed equal or near-equal advantages; fires,
floods, financial depressions, and other disasters could bring an
early and abrupt end to their dreams. But for the moment, all
thoughts and energies were dedicated to the beginning. Even had
they been less confident, and given themselves up to worrying
about what tragedies might defeat their efforts, they still could
not have guessed the cause for the eventual end to lumbering in
the Chippewa valley. In 1857 it simply was not possible to foresee that as the prairies and then the plains attracted ever-increasing numbers of settlers, the demand for lumber would exceed
the supply, and in the process the "inexhaustible" forest would
cease to exist. Measured in years, however, the end was not far
removed from the beginning. It all happened within the life-
times of both Ingram and Kennedy.

CHAPTER III

Hard Times
and a Small Beginning

IN THE UNITED STATES the year 1857 is probably best remembered for its Panic, and in some respects it might have seemed preferable had Alexander Dole, Orrin Ingram, and Donald Kennedy postponed the start of their independent lumbering operation until the advent of more propitious times. But the hard times were not everywhere equally hard, and all businesses were not affected with equal severity. Lumber, for example, continued to be sold, and if prices were lower, so too were the costs of production. This partial immunity enjoyed by lumbermen doubtless was related to the heavy flood of emigration into the trans-Mississippi West in the years immediately preceding the Panic. The demand for lumber continued on for a time, even though the flow of settlement slowed. While it was often possible for the established farmer to delay improvements, more recent arrivals had to provide some protection for family, stock, and grain; and prairie sod was seldom an agreeable alternative to pine boards as the basic material for construction.

In any event, the new partners from Canada initially seemed to pay the Panic little notice. Arriving as it did on the heels of their own western fever, they did not permit the deterioration in financial affairs to alter or even delay their plans. If they were inconvenienced, their own good fortune and the fairly rapid recovery of the economy from the depths of the depression enabled them to survive with little difficulty.

Although prices in the Chicago and Mississippi River lumber markets fell rapidly in the winter of 1857–1858 and would remain depressed throughout the 1858 and 1859 seasons, the firm of Dole, Ingram & Kennedy was not yet in a position to suffer to any

serious degree. The very scope and nature of the operation limited
its vulnerability to such outside factors. During those first few
seasons, the new sawmill between the Chippewa and Half Moon
Lake produced lumber in small amounts and, consequently, any
losses which resulted because of the depressed condition of the
market were minimal. Clearly the extent of their early efforts
could not have been much larger under the most prosperous of
circumstances. Furthermore, since the effects of the depression
were general, it seems likely that Dole, Ingram & Kennedy benefited somewhat from the low prices of the late 1850's, for the
young firm was then buying considerably more than it was selling.

A relatively stable demand with accompanying higher prices
would soon return to the western lumber markets. Coincident
with the improvement in market conditions, the Dole, Ingram
& Kennedy sawmill was prepared to produce increased amounts
of lumber, and the partners were thus able to take advantage of
the better times. While in the midst of their beginning struggles, none of the three could have had the perspective which
would allow an admission that their effort was relatively well
financed. In truth, however, they had capital committed in
sufficient amounts to outlast the Panic and, by the start of the
Civil War, the briefest of glances backward would have provided
some satisfaction.

There were, of course, problems. To a degree the entrepreneurs were hampered in their plans and early improvements because of difficulties in obtaining credit. But the lower prices on
the materials they purchased at least partially compensated for this
lack of credit opportunities. More complex were the problems
which developed within the organization between the partners.
Naturally they did not always agree on plans and procedures, but
too often simple disagreements developed into something more
serious. Soon it was apparent that Dole and Ingram were engaged
in a struggle for dominance. The differences, some of which
could have been avoided altogether by either party, generally
increased in frequency and intensity as the months passed. In
the course of endless debates, basic issues were thoroughly considered and discussed. Under such circumstances, if co-operative
effort was lacking, the inclination to contend at least lessened
the likelihood of any costly mistakes resulting from undue haste.

Unfortunately, however, forced deliberation may have been the sole benefit of this internal struggle.

The widening gulf between Dole and Ingram clearly was not going to be bridged unless one or the other was able to assume a lesser role within the company; but the personality of neither man would permit the acceptance of subordination. Thus the partnership was destined for an early and rather unpleasant end. During the five years of its operation, it was Ingram who generally had his way and Dole who, with some justification, felt used and abused. But these difficulties did not prevent the early establishment of Dole, Ingram & Kennedy on a basis that was relatively secure if not yet prosperous.

Ingram succeeded in taking command largely because he was on the scene in Eau Claire and Dole was not. Although the older partner had stayed behind in Canada, his aid to the Eau Claire effort was of great importance. Like so many young western concerns, the financial roots of Dole, Ingram & Kennedy were eastern and remained so for many years. In this instance, the main artery supplying the working capital for the infant operation ran from Ottawa. There Dole continued to work at his old job, forwarding his earnings from Hamilton Brothers and his borrowings from whatever source to support and sustain the activities along the Chippewa.

In addition to an increase in salary, John Hamilton had offered Dole a loan of $6000 "to purchase timber lands," but the loan was contingent on Dole's promise to remain with Hamilton Brothers for at least five years. In other words, to receive the much-needed capital, Dole had to offer himself as security, with the added assurance that the money loaned would be invested in land. Accordingly the partners had to determine whether, as Dole phrased the question, the money was worth more to Dole, Ingram & Kennedy than Dole's services at Eau Claire.[1] For Ingram and Kennedy, the answer to the question came more easily than the phrasing of their reply to Ottawa. They indicated that the most vital immediate need of the firm was cash; therefore, if Dole could accept such an arrangement, they advised that he maintain his relations with Hamilton Brothers.

---

[1] Dole to Ingram and Kennedy, May 30, 1857.

Thus the senior member of the partnership, at least from the considerations of age and ready means, became to all intents and purposes the outsider. Not that Dole wanted to remain in Ottawa in another's employ. Indeed, the Chippewa valley location had originally been his idea. As he wrote to Ingram on June 10, 1857, the day before the Ingrams arrived at Eau Claire, "I really feel lonesome thinking of the prospect of my remaining here for another year." Perhaps "lonesome" was not the most accurate adjective, but clearly, Alexander Dole would have preferred to be in Eau Claire.

In Canada, in the early days of their acquaintance, Ingram and Dole must have shared considerable respect. They could hardly have joined in such an enterprise had that not been the case. But once separated, any mutual esteem was soon forgotten and their relationship rapidly deteriorated. The tactics adopted in these long-range controversies were as different as the personalities of the participants. Dole tended to be sarcastic, and the deeper his anger and resentment, the more bitter became his sarcasm. Ingram, quick neither of tongue nor of pen, adopted what proved to be an even more effective weapon—silence. In later years, he would demand that his agents write him twice a week whether they had anything of importance to report or not. But in the late 1850's, he provided only the sketchiest information on the progress and the plans at Eau Claire to his absent partner Dole. As a result, Dole was forever begging for news, and his complaint of June 29, 1857, became a very common theme: ". . . must say your letter has given me the Blues. Not so much for what it contains but for what it did not contain."

In these early years of operation, the business records of the company were maintained much like a joint bank account, with careful entry of the amounts deposited by each partner. Dole, the largest depositor, assumed from the beginning that he deserved authority within the organization commensurate with his monetary contribution. Accordingly, he adopted the practice of including directions with his hard-earned and dearly-purchased dollars; but what he intended as directions issued from his desk in Ottawa were received merely as suggestions at the Eau Claire office of Dole, Ingram & Kennedy.

Initially, Ingram tried to give consideration to the profusion of instructions from his older partner. It soon became apparent, however, that what seemed important in Ottawa was often rather unimportant along the Chippewa. Seldom did priorities of the partners coincide, and this was the basic cause of the major differences which developed. It was not long before Ingram viewed the 1300 miles which separated them with gratitude. He expended funds from the company account as he thought best, without bothering to consult his absent partner. Predictably, Dole disagreed with most of the decisions reached. On those infrequent occasions when the old man happened to have no objections, he found it nonetheless disturbing to realize that his Eau Claire partners had made their determination without knowing or even requesting his opinion on the subject. In some respects, their differing attitudes appeared to parallel their earlier experiences in lumbering. Ingram assigned priority to the sawmilling or the production function; Dole sought to expend energies and capital on the acquisition of pine lands, and only partially because of the stipulation in his contract with his employer, John Hamilton. Whatever the reasoning, Ingram's position was decidedly the more appropriate.

The difficulty, if not impossibility, of agreeing on priorities was what might be expected from two men who held very diverse views concerning business affairs. Dole maintained the ideology of mercantilist capitalism. He was apparently intent on forcing the tiny enterprise into an integrated structure, a replica of an old-fashioned firm, secure in its supply and its outlets as well as efficient in the production phase of operation. Ingram insisted on an entirely different approach. He argued that they should be dedicated to specialization of effort at this point in their development, concentrating their very limited resources on producing the best lumber possible at the lowest possible expense. An expansion of activities similar to those then proposed by Dole, including major attention to pine land purchases and marketing procedures, would become appropriate for Chippewa lumbermen in twenty-five or thirty years. In the late 1850's, however, such concerns were far beyond the capabilities of Dole, Ingram & Kennedy.

Without question, the distance separating Dole from Eau Claire was, of itself, an important influence which affected his ideas and altered his proposals. Indeed, his early concern about controlling their own source of supply and later with arrangements for retail sales had to be partially the result of this apartness, which excluded him from the day-to-day details of the business. He tended to be a kind of free spirit, painting in broad strokes, moving from one large and separate subject to another. That his partners chose to keep him uninformed about so many things served only to increase the distance between Eau Claire and Ottawa.

But other differences also existed. Ingram and Dole belonged to different generations. Young Ingram was an engineer of change. Change was as appealing to him as it was distressing to Dole. Ingram was an opportunist and an expansionist. The situation today needed but to hint at the promise of tomorrow, and Ingram assumed that tomorrow conditions would be improved. Dole, on the other hand, moved with caution. His interest in the Chippewa pineries had more to do with considerations of a safe and secure investment than with the possibility of creating an empire. Conservatism in business, as in other areas, is not uncommonly a correlative of age: the young generally have less to be conservative about. In any event, relations between Ingram and the older Dole clearly suffered from the effects of a nineteenth-century generation gap. Many factors contributed, but one of the more obvious involved their assessment of the future.

In a word, Ingram was confident. Dole feared what tomorrow might bring. But the optimism so apparent in the Ingram of this early period cannot merely be ascribed to youthfulness, because it would always remain one of his most important characteristics. He and so many of his more successful contemporaries somehow seemed possessed of eternal confidence in their business outlook. Panics and hard times were passing things, to be endured and in no way to alter or long deter inevitable progress towards wealth and power. This acceptance of expansion and prosperity as the normal course of things was the unchanging backdrop for their activities. Consequently, they were able to hold, for the most part at least, a larger and longer view of business affairs. Failure became but a temporary setback. For men of ability who were willing to work and to make sacrifices, success had to come sooner or later.

Years later, when success had apparently proved the truth of these optimistic inclinations, Ingram was not the least reticent in calling attention to his "bullish" reputation. In the spring of 1890, when demand and prices for lumber could hardly have been better, Ingram replied to a letter from his nephew and long-time business associate William P. Tearse, a letter in which Will had noted a partner's reference to Ingram:[2]

> I am glad to know that he thinks I am disposed to look at the sunny side of things. He is quite correct in that, and I believe if everybody would do the same thing that they would be much happier.

Three years later, after noting the poor conditions for logging operations, Ingram tried to encourage one of his employees:[3]

> . . . the supply [of logs] may be quite equal to the demand. I am looking for that result, but am not losing sleep nights on account of the effect on business. I am inclined to look on the sunny side of all these things. I enjoy myself better in that way, than looking for some unforeseen calamity.

In early 1894, when the pall of financial depression had again settled across the nation, Ingram remained optimistic, or so it appeared from his letters to Hannibal partner, Colonel Dan Dulany:[4]

> I am just in receipt of yours of the 15th inst. with letter enclosed from J. A. Holmes & Co. I see that your ideas and Mr. Holmes are in accord. But I hope that you are both of you too bearish. Holmes I know is a little inclined that way, and you at Hannibal have had times when you sympathized with that crowd; and you know that I am always disposed to look on the sunny side of all things. I agree with you that the out-look for the past few months and the present time is not just what I would like to have seen . . . but still I am inclined to believe that business will turn out this spring fairly well. And if we have a good crop, which you know is the old cry, we shall be very sure to have a good trade. . . .

But for all of his expressions of confidence in the future there were periods when, even for Ingram, optimism was maintained only with difficulty. This was the case as the months slowly passed in 1857.

Throughout the long days following the start of their efforts in Eau Claire, the partners naturally shared many uncertainties. Dole would have been perfectly content had they waited to begin

[2] O. H. Ingram to W. P. Tearse, March 21, 1890.
[3] O. H. Ingram to C. August Staples, March 9, 1893.
[4] O. H. Ingram to Colonel D. M. Dulany, Sr., January 18, 1894.

sawing lumber until the spring of 1858. But both Ingram and Kennedy were anxious to get underway, if only to test the machinery and procedures to insure that all was in readiness for a full season of production the coming year. Their work that first fall ended sooner than they had expected, the early arrival of the Wisconsin winter taking them by surprise. Only a little lumber had been sawed. On the portable mill purchased from Randall, however, they did succeed in cutting out the boards and timbers needed for the completion of their own permanent sawmill as well as one for their neighbor, Daniel Shaw.

There was no lack of causes for worry among the valley lumbermen as they prepared for the start of logging in the Chippewa pineries. They had begun to think more about cutting losses and less about turning profits. Although Dole, Ingram & Kennedy still had no lumber ready for sale, this was consolation of a small order. Indeed, Ingram and Kennedy were exceedingly anxious to get underway in all phases of the operation, even if this meant selling their lumber at low rates and on unfavorable terms.

The reason for their impatience was clear enough. The commencement of any business activity usually involves a long delay from the initial investment to the receipt of the first returns on that investment. During this period, funds are expended and debts incurred with the promise of any income still far removed. Anxieties increase as resources decrease, and when credit opportunities all but disappear, the chances for success are further reduced. Dole, Ingram, and Kennedy could not escape worrying whether their limited financial resources could outlast payments until the time when earnings finally became a factor.

For the most part, however, they tried to emphasize the more positive aspects of their situation. The partners were correct in believing that in some respects it was a relatively good period in which to begin operations. Good times or bad, they would not have had much money at the start, and their few dollars went further in 1857 than they would have gone during more prosperous times. Furthermore, they assumed that by the time they had lumber for sale, conditions in the markets would be improved. Another favored method of combatting their own personal depression was by enunciating a theory that panics had a leveling effect—that older and established firms were so weakened by hard times that younger and less substantial organizations, such as their own, were better able to compete.

True or not, such possibilities failed to alter the financial requirements of Dole, Ingram & Kennedy. The new firm had but little cash and little credit, and little hope of soon obtaining much more of either. In addition, during such times the establishments with whom they did business could not allow terms which might have alleviated their hard-pressed situation.

Currency was clearly their greatest immediate need. Even during so-called good times, in frontier communities there seemed never to be enough cash available to conduct normal business affairs. In the course of depressions, cash all but disappeared from western markets. During the fall of 1857, admitting that the financial situation was bad and growing worse, Dole suggested only half in jest, "it would be well during the present time to have an Indian currency, that is a *Beaver skin to be worth so many Martin* & so many skins worth a blanket."[5]

But Indians they were not. They needed cash to pay for logs, machinery, and the other materials and wages necessary for production, not to mention funds needed to satisfy Dole's obsession with pine land purchases. Ingram and Kennedy fully appreciated that the wellsprings of their credit were very shallow and that the firm had to become self-sufficient in a very short time if it were to survive. There was only one way to achieve this self-sufficiency, and that was through the production and sale of lumber. In the fall of 1857, however, proceeds from their own lumber still seemed a distant objective. As Dole reminded his younger associates, in a moment of optimistic weakness, "All undertakings are attended with more or less difficulty but think we may yet realize the advantages of our location."[6] It was a time for enduring.

Almost from the beginning, the overriding need for cash forced the partners to think in terms of turning logs into dollars, even on conditions dictated by the buyer. That their situation was not unusual among western lumbering firms only served further to depress the market. In any event, Dole, Ingram, and Kennedy

---

[5] Dole to Ingram and Kennedy, October 8, 1857.
[6] Dole to Ingram and Kennedy, January 9, 1858.

learned, although not quite simultaneously, that losses are relative and there are times when necessity demands that goods be converted into cash at almost any price.

Efforts to obtain operating capital were complicated by Dole's insistence that they immediately begin acquiring pine land. Such expenditures could only aggravate their already difficult circumstances, but Ingram was unable to convince Dole of this. The distant partner was certain that it was the ideal time to make such investments because, as he reasoned from Ottawa, prices on timber land were then depressed to the point which allowed those with funds the opportunity to take advantage of exceptional bargains. In addition, Dole argued that they ought to be making their timber selections while they had such a free choice, a situation he feared would not long continue. Either through underestimating the amount of standing pine in the valley or overestimating the number of farsighted lumbermen like himself, at this early date such fears were premature. Nevertheless, when Mrs. Kennedy left Ottawa in late September 1857 to join her husband in Eau Claire, she carried with her a package containing $2000 from Alexander Dole, the money to be used "for Land Purchases."[7]

Ingram naturally preferred greater latitude in the expenditure of funds, and in this instance he requested additional directions from Ottawa, informing Dole that the land office in Eau Claire had not yet even opened. In his reply, Dole expressed some regret for having sent the money, admitting that he had assumed land sales were being made in the district. Nevertheless, he declined to alter his original directions, insisting that the money be "applied in the manner intended." Dole assumed the land office in Eau Claire would soon be ready for business. If not, he suggested the possibility of arranging trades with private parties, providing the land could "be obtained at prices near that of the Government."[8]

The man with the means cautioned, however, that his partners be especially careful "about having clean titles as hard times gives some people hard faces and still harder consciences." It seems unlikely that Dole's advice was well-received by Ingram,

[7] Dole to Ingram and Kennedy, September 26, 1857.
[8] Dole to Ingram and Kennedy, November 11, 1857.

for not only was the money needed to meet the costs of current operations, but also the admonition concerning land titles came too late and probably generated additional hard feelings.

Ingram had recently learned from experience the dangers of closing a land purchase without due regard to title. Their own sawmill property, which they thought they had bought from Adin Randall, the enterprising real estate dealer of West Eau Claire, was discovered to have a mortgage on it. The resulting legal difficulties would trouble the partners for many years to come. Nearly a decade later they were still attempting to untangle the confused puzzle which Randall had created. With the objective of finally gaining clear title to their mill site property, Kennedy returned to Ottawa during the winter of 1866–1867 to discuss the matter with Dole, who was by that time no longer a partner. Disgusted with all they had been forced to go through, the usually mild Kennedy advised Ingram that never again would he have anything to do with a piece of land on which Randall had any prior claim, "as I would be sure that there would be some judgment or mortgage or something else to annoy as long as a person lives."[9]

Thus, for good reason, Adin Randall headed the Dole, Ingram & Kennedy list of those to be avoided in future business dealings. Even Dole had observed from his distant perch, "I have no confidence in Randall whatever," adding that it would surely be a mistake to lend him anything without security, "and even security itself with him is doubtful [since] he is such a man to mortgage, etc."[10] (This apparent dishonesty seems to have been eventually forgiven if not forgotten, and years later Ingram donated the statue of Adin Randall which still stands in little Randall Park on the west side of Eau Claire.)

In late 1857 the conflict over the sawmill property title came

[9] Kennedy to Ingram, December 27, 1866. Adin Randall must have been quite the character. The *Eau Claire Free Press* of February 17, 1859, in calling attention to an advertisement run by Randall, observed that "Mr. R. is 'up and doing,' and as usual is prepared to buy, sell or trade in anything that ever was invented." The text of Randall's advertisement read: "I am prepared to sell business and residence Lots, Farming or Pine Lands, which I offer upon the most reasonable terms, or receive in exchange labor, oxen, horses, lumber, logs, or personal property of whatever name or nature. Call and see if I don't trade with you."

[10] Dole to Ingram and Kennedy, November 11, 1857.

close to being the last straw so far as Ingram was concerned. For all of his confidence and optimism, he nearly packed up and returned to the certainties of employment in the Ottawa valley. Had it not been for the enthusiasm of Adin Randall, Eau Claire's indomitable town-boomer, Orrin Ingram might have given up on a future of lumbering in the Chippewa valley.[11] In retrospect, Randall's duplicity dimmed in importance, and a wealthy and successful Ingram confirmed that although Randall may not always have been above reproach in his business dealings, he was, at least, correct in predicting that investment in the Chippewa pineries would prove rewarding.

After reaching the difficult decision to remain in Eau Claire, Ingram recommitted himself to the task of building a plant and an organization which could produce lumber for sale at a profit, even if prices remained low. In the process, he decided that he could no longer pay attention to the suggestions or directions offered by partner Dole. They may have agreed on the subject of land titles and on Adin Randall in 1857, but they agreed on little else and certainly not about pine land purchases. There was no possibility of convincing Ingram that timber investment deserved any serious consideration, much less that any of their funds should be allocated for such purposes. In the Chippewa valley there seemed to be an abundance of pine available and almost no ready cash.

This disagreement between Dole and Ingram soon became a basic point of contention and caused increasing friction as the months passed. In part, Dole was committed to timber purchases by the arrangement he had made with his Ottawa employer— but only in part. Ingram suspected that Dole overemphasized the terms of his employment with the Hamiltons in order to further his own ideas concerning the priorities for Dole, Ingram & Kennedy. Accordingly, in the spring of 1858, Ingram wrote to Dole requesting that he attempt to borrow some money from the Hamiltons for the specific purpose of improving the sawmill. Ingram also bluntly complained that he did not appreciate receiving money which had so many strings attached that it could not be spent. The Eau Claire land office still had not opened.

---

[11] See, for example, Bailey, ed., *History of Eau Claire*, 841–842.

But Dole remained insistent that the funds be used as directed.[12] It was probably fortunate for the future of the firm that Ingram had decided to accept the advice from Ottawa only when convenient, and to spend the money where needed.

Whether or not Dole, Ingram & Kennedy owned its own pine lands, one of the unavoidable decisions facing the partners as they looked towards their first full season of operation was how many sawlogs to purchase. They had received a good offer for some 350,000 feet of logs. But Dole, who now seemed to find planning and preparation nearly an end in themselves, would not approve such an investment. He advised that any decision to purchase logs should be based on "the prospect of sales for lumber and the stock on hand by other parties, *not sold*," and since the immediate outlook for sales was at best uncertain, he did not see how they could consider closing the deal. Dole also took this opportunity to remind his Eau Claire partners that "it is no profit to us with our capital to hold logs over not cut into boards, as we have this winter."[13] Clearly the old gentleman was still annoyed that, against his wishes, Ingram and Kennedy had started up the mill the previous fall and had sawed a small amount of lumber.

There was, of course, some basis for Dole's caution. Just as clearly, however, there could be no profit for the firm until some lumber was produced; and, good times or bad, the chance had to be taken. Their limited resources made it impossible to wait out the depression. If, as Dole warned, uncut logs carried no profit, neither did an idle plant.

Thus, in the spring of 1858, Dole, Ingram & Kennedy began to saw lumber for the market. The Eau Claire partners knew far more about making lumber than they did about the market. But since the quantity produced was initially too small to warrant the retailing or even the wholesaling of their own produce, Ingram arranged to sell some of the earliest manufacture to two of the better-established lumbering firms in the valley: Carson, Eaton & Rand of Eau Galle, and Chapman, Thorp & Company of Eau Claire. Dole was predictably much displeased. From Ottawa, he charged Ingram with neglecting to consider the ex-

---

[12] Dole to Ingram and Kennedy, April 17, 1858.
[13] Dole to Ingram and Kennedy, May 25, 1858.

pense of running the lumber down river, which Dole estimated to be three dollars per thousand feet. Had Ingram been aware of this, and he should have been, Dole could not understand how he could possibly have agreed to the price of ten dollars per thousand. He concluded his remarks by observing that selling at such a low figure "seems to me worse than putting the lumber in piles on the Bank."[14]

In a subsequent trade, Ingram arranged for sale of their lumber at eleven dollars per thousand feet, but Dole refused to be encouraged.[15]

> [I] think your lumber if even only ordinary [he wrote] should be worth something more than common prices from being better sawed, and I can only state as an opinion that the lumber would be better piled at [the] mill than at St. Louis at that price could we manage to do so, and prefer to saw with one saw this year again, in preference to giving in to a cramped and complicated, embarrassed business with claims unsettled and expenses accumulating that markets did not justify.

Although the lumber may have appeared underpriced from the vantage of Ottawa, Ingram had managed to get about what it was worth in the Mississippi River markets.

One of the reasons the partners were drifting further apart was Dole's lack of understanding concerning market conditions in the west, and how that market operated. Lumber produced in the mills along the Chippewa in 1858 was floated to market on the rivers. There was as yet no rail option. The character of the rivers, of course, varied with the seasons, but the depth of the water could also vary from day to day. It was not always at a stage which permitted the movement of the lumber rafts to the downriver markets. Thus delivery, already made seasonal by winter ice, was often further complicated and restricted by summer drought. Practically, this meant that lumber arrived at Mississippi River levees in an irregular pattern, and when it did arrive, it came in large amounts. It was not uncommon for many rafts from many companies to be competing for the buyer's attention and his purse. Whatever the total amount of lumber available to a market throughout the season might be, on the days of sale there always seemed to be a surplus.

[14] Dole to Ingram and Kennedy, June 19, 1858.
[15] Dole to Ingram and Kennedy, June 24, 1858.

During this early period, small producers like Ingram had to auction their lumber, lath, and shingles at towns along the river. Firms in dire need of funds were naturally subjected to the most severe bargaining by knowledgeable buyers. But each sold to whomever they assumed had made them the highest bid. As the raft moved downstream, offered for sale at each successive point, the owner had to decide if the current offer was the best he would receive. Should he sell now or move on, taking a chance on the possibility of a higher bid downstream? If he refused a trade and something better did not materialize, he could not return. For logs and lumber, rivers were one-way streets. The seller often had second thoughts, but he did not get second chances.

So Ingram and Kennedy, able to sell but a portion of the mill product to Carson, Eaton & Rand and to Chapman, Thorp & Company, faced the uncertainties of having to dispose of the balance of their lumber somewhere downriver. Their almost complete lack of knowledge of the downriver markets might well have proved a critical deficiency, and without a bit of luck they might have ended the season of 1858 with considerable lumber yet unsold.

The good fortune actually began some months earlier, when Ingram negotiated a contract with E. D. Rand & Company of Burlington, Iowa, for some logs they had left behind on the Eau Claire River at the time of the sale of their property on that Chippewa tributary to Chapman, Thorp & Company. The terms of this agreement provided simply that in exchange for the logs, Dole, Ingram & Kennedy would deliver to E. D. Rand & Company at Burlington one-half the lumber manufactured from those logs. This was an ideal arrangement for the times, particularly for Dole, Ingram & Kennedy, because it involved no exchange of cash.[16]

As agreed, the lumber reached Burlington in the summer of 1858, but in the process Dole, Ingram & Kennedy approached the brink of financial disaster. Although their debt to E. D. Rand & Company had been settled by the delivery of the lumber, no cash resulted and half a raft remained unsold. Unless some

[16] Ingram made a similar trade with Samuel F. Weston of La Crosse, Wisconsin, in the spring of 1859. See Weston to Ingram, April 30, May 7, and August 10, 1859.

early disposition of the balance of the lumber could be arranged, a raft crew would be unpaid. Such a prospect was disquieting for a number of good reasons.

But the worst did not occur. Rand was sufficiently pleased with the quality of the lumber received that he offered a letter to Ingram to be delivered to John Whitehill, a St. Louis dealer. In the letter Rand suggested that Whitehill consider purchasing lumber from Dole, Ingram & Kennedy, assuring him that he could do no better. Years later, Ingram acknowledged: "Without the letter I do not know what I would have done." In any event, Whitehill was willing to advance the money needed to pay off the raft crew, and, of equal importance for the future at least, another contact in the market had been established.[17] There would be many critical trades in future seasons, but probably none would be more important than the 1858 trade with John Whitehill.

These early successes in the market placed Ingram in a far better position to face his partner Dole. Finally arranging a leave from Hamilton Brothers, Dole came west in late summer to see for himself just how the little firm was getting along. There is no question that he desired to make a positive contribution to Dole, Ingram & Kennedy. Although he visited Eau Claire, the Ottawa partner spent most of his "vacation" in St. Louis, attempting to locate buyers for their lumber. Times continued hard, however, and the task of finding buyers for anything, even quality lumber, was discouraging.

Nevertheless, the trip was obviously of benefit. His personal inspection convinced Dole that Ingram was not doing such a bad job after all. Indeed, in the course of his return journey to Ottawa, Dole addressed his partners from Chicago in warm and friendly terms: "I am happy to say that my trip to the Chippewa has not tended to decrease my confidence in your good management and trust by the Blessing of a kind providence we may prosper in our business." He concluded his letter "With best wishes and congratulations to Mr. and Mrs. Ingram on the auspicious occasion of an addition to their family circle."[18] The Ingrams' first-born child had died in New York prior to their move

[17] Ingram, *Autobiography*, 48.
[18] Dole to Ingram and Kennedy, September 23, 1858.

west. Dole's reference was to a son, Charles, born in Eau Claire on September 12, 1858.

But the arrival of Charles Ingram and the departure of Alexander Dole provided but a brief period of pleasure and peace. The concerns of business were dominant. It seemed for a while in that first full season of sawing that their expertise might prove insufficient in establishing a successful operation. Like many struggling lumbermen of that day, they sought opportunities other than selling lumber to provide additional cash. A common effort on the part of sawmill owners was to construct a grist mill which, since it could use power already available but unused during the winter months, involved no great expense. Beginning in January, 1858, Dole, Ingram & Kennedy advertised their readiness to mill flour at reasonable rates. The response, however, was less than had been hoped, and the supplier of the machinery for the grist mill, John T. Noye & Company of Chicago, impatiently joined the list of those awaiting payment—all the while charging 12 per cent interest.

With the approval of Dole, and with the same objective of attracting cash, Ingram launched another project. Near the sawmill and the grist mill they constructed a little general store, known in later and more successful years as Ingram, Kennedy & Mason. Initially they called it a "lumber store," and they planned to keep in stock only those articles and goods necessary to their lumbering activities, particularly as concerned the logging operations. Thus they kept a supply of socks, boots, mittens, blankets, molasses, salt, pork, and like items. In the beginning, Dole had voiced no objection to the plans. Most of the articles were needed by the firm anyway and might just as well be purchased wholesale; and if they could acquire some cash by sales either to their own loggers and mill employees or to anyone else, so much the better.

But the store, like the grist mill, proved a financial disappointment, and Dole reconsidered his earlier approval. Attempting to assign some blame for the poor showing of the store, Dole inquired of Kennedy about the truth of a rumor which had reached him in Ottawa—that the ladies of Eau Claire avoided the store out of fear of Ingram's "amorous propensities."[19] This inference

[19] Dole to Kennedy, October 1, 1859.

was absurd, as Dole must have known. It was completely out of character from every consideration; and, most importantly, in their tiny store in that small town, a man so in need of dollars could not permit distractions. Under such circumstances, Ingram would far rather sell a skirt than see one.

In the summer of 1860, when Dole visited the store, he examined its inventory and books and uncovered what seemed to him a number of blunders. In his caustic way, he took particular note of a "lot of Mens Prunella gaiters . . . [which] cost $2.00 in St. Louis and will likely remain on the Shelves as a memento to profit & Loss, along with some other similar articles, Being about as much needed and as appropriate for the sands of Eau Claire as crinoline for the Sioux Indians."[20] Indeed, times were hard.

More than Ingram could realize, all of this was good training. The lack of operating capital, assumed to be largely the result of the depression, would prove to be a chronic deficiency. In a sense, Ingram learned to thrive under conditions best described as hard-pressed. He came to believe that a business with plenty of available capital ran the danger of avoiding healthy pressures that forced the decisions necessary to efficiency and growth. Regardless of his subsequent success, Ingram would not rest and would never allow cash to lie idle merely for the purpose of settling a bill nearly due. So a shortage of funds was a way of life and not just the result of financial circumstances outside the company office. This did not make the effects of the Panic of 1857 any more agreeable to Ingram or his partners.

Dole, Ingram & Kennedy were severly constrained by the depresion of the late 1850's and were afforded little opportunity to defer some basic decisions. But, as was the case with the selection of the mill site, it is doubtful that more time and less pressure would have produced wiser policy. Certainly Ingram was prepared to make the decisions appropriate to their situation and his own ambitions. They had already managed to construct a fine little steam mill, worth an estimated $10,000, on an ideal location between Half Moon Lake and the Chippewa. With a crew of twenty men, they could saw good lumber as economically as any mill on the river.[21] During the 1858 season, the capacity

---

[20] Dole to Ingram, August 20, 1860.
[21] *Eau Claire Free Press*, September 23, 1858.

of the sawmill was about 15,000 feet of lumber per day, operating one muley-saw and two gangs, one of eighteen and one of ten saws—perhaps the first modern gang saws ever used in Wisconsin.[22] Mounted in parallel fashion in a single frame, these gang saws revolutionized the mill operation. It was an amazing spectacle for those familiar only with former methods "to see them take a log about the diameter of a sugar hogshead and twenty feet in length, and by passing it through a gang once, reduce it to a pile of inch boards."[23]

The following winter Ingram and Kennedy installed a new engine which increased the power and sawing capacity of the mill by nearly nearly two-thirds, enabling them to handle upwards of a hundred logs a day, or some 25,000 feet.[24] Clearly, for all of the early difficulties, the partners planned on better times and prepared themselves to sell some lumber at a profit. But again they were disappointed.

Although the general financial situation appeared to have changed very slightly if at all, as the partners looked forward to the 1859 season they realized that their own situation was considerably stronger than it had been the year before. In the first place, with the experience acquired and the contacts they had established in the markets, they were confident that their lumber would be able to compete with any on the river. The prospect of continued low prices seemed to cause them little concern for, as they noted, "Our logs cost us about $3.50 per M [thousand feet] this year, so we can afford to sell [lumber] low."[25] They would not be alone in underestimating just how low "low" would be.

The season of 1859 began inauspiciously even for a depression year. The highest water experienced on the Chippewa since 1847 delayed the start of sawing and also resulted in the loss of some logs when a boom that Dole, Ingram & Kennedy shared with Adin Randall broke.[26] Continued high water on the Mississippi de-

[22] See the biographical sketch of Ingram in *American Lumbermen.*

[23] *Eau Claire Free Press,* June 21, 1860.

[24] Dole to Ingram, June 15, 1859.

[25] Dole, Ingram & Kennedy to Bagley & Sewall, May 18, 1859.

[26] The *Eau Claire Free Press* of June 2, 1859, estimated that some 600,000 feet of logs were lost as a result of the high water, which was doubtless an exaggeration. Still, the loss was considerable and occurred at a most inopportune time. See *Milwaukee Sentinel,* June 8, 1859.

layed the movement of the lumber downriver. It was June before Captain Jeremiah Turner, who would pilot rafts for Ingram many seasons, was able to leave Reads Landing, Minnesota, with the first raft for the downstream markets.

So far as possible, the partners shared the marketing responsibilities during these first years of operation. This was wise policy for a number of reasons, most importantly because it insured that each gained some appreciation of the difficulties involved, thereby lessening the tendency to complain about the prices accepted by another. When Dole was able to arrange for time away from Hamilton Brothers, he continued to spend most of his days in St. Louis and other river ports, attempting to solicit future sales and to collect on past trades. At other times Kennedy or Ingram would go downriver, keeping a town or two ahead of their raft, looking for likely buyers. By this procedure, the raft would not have to go through the difficult and time-consuming task of stopping unless signalled to do so because a sale had been made.

But too often signals were few and the rafts, all or in part unsold, would reach St. Louis, which was for practical purposes the end of the line. During the summer of 1859, the partners tried to hold their lumber to prices averaging no less than twelve dollars per thousand feet at St. Louis and correspondingly lesser amounts at points farther north, closer to Eau Claire. Although their boards were generally of high quality, sound and evenly sawed, it remained difficult to find interested buyers with cash. Ruling prices commonly ranged at less than ten dollars and on some occasions, when buyers traded with those having small lots who were desperate to sell, prices fell below nine dollars per thousand feet.[27]

Market conditions had not improved by the fall, and a discouraged Alexander Dole observed that there was but "little chance of making a contract for lumber at present as every place seems to be full." Recently he had watched Chapman and Thorp "retailing their last raft out on the Levee" and, he warned his partners in Eau Claire, "it is very probable we will have to do the same."[28]

[27] Jotham Clark, McGregor, Iowa, to Ingram, June 20, 1859.
[28] Dole to Ingram and Kennedy, September 5, 1859.

Dole was beginning to understand that Ingram was not entirely to blame for their problems marketing lumber. Still, it seemed there had to be a better method of doing business, and Dole spent much of his time considering alternatives to forced sales, often made in direct competition with larger and stronger organizations. The most obvious solution involved the establishment of a permanent yard at some downriver point where their lumber might be piled and held. The mere existence of such an option would permit the seller to bargain with greater strength in the markets. But obvious and logical as this solution appeared to be, like so many simple answers it was attended by more difficulties than were initially foreseen. Whether practical or not, a downriver retail yard became the new focus of Dole's interest and attention, for a time replacing even his preoccupation with pine land purchases. Convinced almost immediately that a downriver yard was the solution to their marketing problems, the next step as viewed by Dole was the selection of the best possible location.

His enthusiasm was evident as he addressed Ingram and Kennedy on the subject: "the question is would it not answer well for us to establish yards here next spring by landing at Hannibal, sell from a yard there and have also one at some point on the [rail] road as soon as one knows who to sell to."[29] Clearly his plans had already expanded beyond a single yard to several, radiating out from a wholesale depot in Hannibal.

In all likelihood, Ingram studied Dole's ideas with considerable misgivings, as an indication that his partner had veered off on a new tangent. Ingram felt uneasy for good reason. Dole was beginning to regard his lumber yards as crucial to the company's operation, possibly of even greater importance than the customary wholesale contracts made directly from the water. In addition, the Ottawa partner soon considered the establishment of lumber yards to be his personal contribution to the eventual prosperity of Dole, Ingram & Kennedy.

Dole explained his preference for Hannibal not in terms of the town itself but rather in terms of the region to the west. He described the area between Hannibal and St. Joseph as a "new and a beautiful farming country, [and] people generally have small capital and a good chance. . . ." There were other reasons for

---

[29] Dole to Ingram and Kennedy, September 7, 1859.

selecting a yard site above St. Louis and the confluence of the Missouri River. Dole considered "the landing at Hannibal as much safer, easier, and Boards can be taken out cleaner than at St. Louis where all is unpleasant and property is at more risk . . . and every thing done at high prices."[30]

Although Dole managed to acquire a good deal of information about Mississippi River lumber markets, his western education was too limited to be of much benefit to Dole, Ingram & Kennedy. He had vaulted from land buying to retail yards, altogether ignoring the intervening problems of processing. Ingram may have appreciated his partner's efforts, but he could hardly approve the resulting recommendations. Dole's plans were simply too large and impractical to be given serious consideration by the young firm in 1860. The day would come when the management of a downriver wholesale yard was the only way to do business, and the particular advantages of Hannibal would again be recognized; but that was much later. In the meantime, if their own inexperience in sales was not sufficient to bring a halt to such big ideas, their lack of capital was.

Ingram realized, if Dole did not, that attempting to sell strictly from their own yards was a luxury they could not yet afford because the cash returns from such sales were uncertain and slow at best. Although the possession of a yard or yards downriver might lessen the necessity for selling at unfavorable terms, it could not lessen their need for cash. Despite the possible advantages of a Hannibal location, at this point in their development St. Louis was the preferable site for a yard, if only because it was farther downriver. Of necessity, the primary consideration was to sell as much lumber as quickly as possible and to pile only what remained unsold. Therefore, the maximum opportunity for sales from the water had to be provided, and accordingly the yard had to be located at "the end of the line." Battle lines were promptly drawn between Dole and Ingram on a new subject. The older partner was convinced, even at this early date, that downriver yards were vital to the success of the company; Ingram, that such yards in 1860 were little more than a dumping ground for leftover lumber.

For all of his convictions, and, one suspects, almost in spite

---

[30] Dole to Ingram and Kennedy, September 7, 1859.

of himself, Dole did succeed in negotiating some contracts in the fall of 1859, one of which was for 400,000 feet of lumber. In this instance, however, Ingram chose to veto his partner's arrangements, stating that the ten-dollar price was too low and the terms—one-third cash, one-third in three months, and the balance in six months with no mention of interest—too unfavorable. It had become nearly standard procedure for Dole to complain about contracts negotiated by Ingram. Now Ingram took his turn.

Yet it was this ability to obtain contracts on their lumber which made the subject of establishing their own wholesale-retail yard less critical. Dole realized this. In anticipation of just such reasoning by Ingram, when Dole forwarded the terms of the contracts to Eau Claire for consideration, he took care to add: "I hope you will not think that I have given up the Idea of yarding our lumber." On the contrary, he informed his partners that he was in complete agreement with the attitudes so commonly expressed by sawmill owner-operators, and was now firmly convinced of the absolute necessity of a lumber yard. Accordingly, Dole requested that Ingram and Kennedy "make every future arrangement and expenditure of minor importance to this one object." He wrote:[31]

> I believe there is very few in the trade but would screw down the Manufacturer to the last dime of the actual cost of the lumber delivered to them and that they make their calculations to keep the manufacturer down if possible and rather encourage anything that will keep him in debt, subservient to his money and power which makes it absolutely necessary for us to streatch every point to check all outlay which may compell us to such Immediate sales to meet and begin to yard our lumber as soon as possible. the sooner the more profit we will make of it— before going up [to Eau Claire] I expect to prepare some proposition for yarding either this fall or spring as our circumstances will permit. You must see by my last letter that we get no share of the profits made on our lumber, and every Lawyer, Quack or idler who has credit or a few dollars to spare to dabble in a lumber yard is making more profits out of it than we are.

Like many other lumber producers of the period, Dole readily accepted the popular notion that the middleman—the hated wholesale dealer who on the surface appeared to contribute so

[31] Dole to Ingram and Kennedy, September 10, 1859.

little and take so much—was the real source of their difficulties. It followed that if the mill owners could manage without the wholesaler, their own profits would increase substantially. Although appealing, such a theory was defective in some important particulars, as those who attempted to assume the wholesaling function quickly learned.

By eliminating the wholesale dealer, producers would enter the new and unfamiliar field of marketing. Ingram knew that they and other small operators were neither in a position to compete for retailers over a widespread area, nor able to accept the additional delays in return on investment which such an extension of their activities imposed. Also, despite his reputation among producers, the wholesaler was considerably more than a mere distributor of lumber; and if the millmen were intent on replacing him, then they in turn would have to do the work which he had formerly done. Primarily, this involved preparing the lumber for sale to the actual user. As middlemen, the producers would have to provide for drying or curing the lumber by means of dry kilns or extensive piling grounds, either of which cost money. In addition, it was expedient if not essential to have access to a planing mill where rough lumber could be finished. Finally there were arrangements to be made with railroads for delivery, a responsibility which required constant attention as to freight rates and availability of cars.[32]

Some of these factors were as yet unconsidered by either Dole or Ingram, and therefore had nothing to do with the basis of their differences. Had the resources of the company been larger, the disagreements between Dole and Ingram might have been alleviated to an extent by allocating some funds toward the priorities of each partner. In other words, with a larger pie to divide, there would have been greater opportunity for compromise. Such was not the case, however, as they prepared for the 1860 season, and when Ingram proposed that any and all available resources be directed towards the improvement of the sawmill he found Dole in determined opposition. Ingram contended that there would always be a market for lumber of good quality and at this stage in their development, they did not possess the means to involve themselves in both production and marketing.

[32] See the *Northwestern Lumberman,* November 9, 1878.

Making lumber was their profession; it was what they did best. Dole countered by arguing that Ingram had gotten things turned around, that "if expenditure is to be made in improvements, machinery and other depts. not actually necessary, so as to compel forced sales to meet them, we will never find ourselves able." Immediate self-sufficiency underlay his line of reasoning. While willing to admit that the mill improvements advocated by Ingram were desirable, Dole argued that they should be delayed until "the proceeds are raised [rather] than to embarrass sales for the sake of them before hand." He also thought it surprising that recent experiences had not convinced Ingram of the folly of continuing his policy. Indeed, Dole felt it should be unnecessary to remind his young associate of their predicament. Business had been bad from the beginning, and "When interest is doubling on the one hand and forced sales on the other at a loss, it is like settling a Bankrupt Estate."[33]

Discussion of the issues among the owners did not seem to serve much purpose. But since Dole's annual leave from his Ottawa employment was nearly over, the resident partners were content to bide their time, await his departure, and thereafter pay his wishes little heed. Obviously the old gentleman would not return home with the same warm and friendly feelings of the previous year. On the contrary, during the last few days he spent in St. Louis that fall of 1859, Dole's grievances against Ingram increased considerably and he made less and less effort to keep them to himself. Not only did he complain about the condition of the lumber, which he found to be a poor product, either too thin or too thick, uneven, and too often bearing evidence of bad judgment in the determination of what kind of lumber could best be sawed out of each log; but also the management of Dole, Ingram & Kennedy generally displeased him. In vexation, he wrote toward the end of his 1859 sojourn in St. Louis, "[I] am at a loss to know whether I should go to Eau Claire or not as it appears the Books are not yet in a shape to show the state of the business . . . which shows rather a deplorable state of things for so short a time."[34]

There was growing suspicion between the partners that their

[33] Dole to Ingram and Kennedy, September 14, 1859.
[34] Dole to Ingram and Kennedy, September 26, 1859.

differences were fast becoming more serious than mere disagreements. Mutual trust had become a forgotten virtue. Despite the disadvantages of being the absentee partner, Dole was unwilling to surrender without a struggle. As a parting shot, just prior to leaving for Ottawa, he made arrangements to lease a small lot near the Ohio & Mississippi Railroad station in East St. Louis, a lot where he piled the lumber rafted during the 1859 season that he had not been able to sell. Although he later admitted that possibly some of the lumber could have been sold for $10 or $10.50 per thousand feet, he reasoned that such prices were ruinous and to trade on those terms would have been nothing more than "doing business for others. . . ."[35] In any event, Dole had his lumber yard.

Following the departure of Dole for Ottawa, Donald Kennedy went downriver to take his turn assisting with the disposal of lumber piled at East St. Louis and with any rafts yet to arrive. Unlike his partners, Kennedy was not committed to any position concerning the propriety of yarding operations, but soon after his arrival in St. Louis, he began to express doubt that Dole had gone about things correctly. In a letter to Ingram written in late October, Kennedy remarked, "if he [Dole] had stayed here much longer he would have got the lumber scattered all over the state of Illinois." Not only was there the yard near the Ohio & Mississippi Railroad depot, which they had known about in Eau Claire, but there was also "an agency at Venice [and] he proposed sending some out to O'Fallon." Kennedy observed, with an almost perceptible shrug, "I don't know but this may be all well enough, but it looks to me like scattering a little business over a good deal of ground."[36] (Dole, back in Ottawa, recalled that he had gotten their retail operations off to a good start, but he did acknowledge: "the *trouble* was in getting the pay. . . ."[37])

Kennedy's reports from St. Louis increased Ingram's suspicions that Dole should not be trusted. Kennedy was beginning to experience some similar doubts himself. The more he saw in St. Louis, the more he became convinced that Dole had not been altogether conscientious in his efforts to sell the rafts in the

[35] Dole to Ingram and Kennedy, October 8, 1859.
[36] Kennedy to Ingram, October 24, 1859.
[37] Dole to Ingram and Kennedy, October 25, 1859.

water. Instead, Kennedy concluded that Dole had been more intent on piling the lumber in a yard where it could be held over until the following season, pending an increase in demand and higher prices. Certainly they all looked forward to better days and bigger profits, but the question posed by Kennedy—"How are we to get along in the meantime?"—could hardly be ignored. Furthermore, Kennedy added with more realism than Dole had yet been able to muster, "I don't think we have any right to expect to make money very fast such times as these. I think that if we can get $12 for it right through we had better take it . . . especially when we can get the cash for it."[38]

But even these expectations were to be disappointed and Kennedy, anything but a traveling man by nature, suffered through the long winter of 1859–1860, making few sales in the St. Louis market. The fault lay not with him, for his efforts ought to have sold more lumber than had been piled in the little yard, nor with his partners, whose suggestions arrived almost too frequently. "I have had a letter from Mr. Dole," Kennedy advised Ingram, and "as usual he has a great many propositions about the different points for carrying on business."[39] During that period, however, there simply was not much lumber of any make being sold by anyone in any of the Mississippi River markets.

Kennedy's greatest frustration throughout his winter stay in St. Louis resulted from the difficulty he experienced in obtaining railway cars to deliver the lumber to customers. It was bad enough to make no sale for lack of a buyer, but it was somehow much worse to lose a buyer for lack of a car. Although he made many attempts to obtain some redress from the various railway officials, he found them uniformly to be "as independant as Lords." In January, while limited to the use of a single car per day, Kennedy offered what might serve as a standard complaint against all bureaucracies, anywhere and at any time, observing that there were "so many officials connected with these railroads and if any thing is wrong each one blames the other, and the superintendant blames the whole so that amongst them they do just as they please."[40]

[38] Kennedy to Ingram, October 29, 1859.
[39] Kennedy to Ingram, November 28, 1859.
[40] Kennedy to Ingram, December 24, 1859 and January 4, 1860.

Kennedy could not then realize that problems resulting from an insufficient supply of railway cars were to prove a chronic aggravation to lumber dealers. He knew only that the railroads were complicating an already difficult task of selling lumber. But in spite of his lack of success in St. Louis and his homesickness for Eau Claire, he resolved "to stay till I can get some money out of it if I should have to stay till next April," because, as he wrote to his friend Ingram, "I know that you need it very badly up there."[41]

During this period, the partners operated under constant pressure and were consequently more sensitive than was normally their nature. Even between Ingram and Kennedy the bonds of friendship proved briefly unequal to the strain—a most unusual occurrence in their long business association. Ingram, in Eau Claire, was finding it increasingly difficult to see beyond the mounting pile of unpaid bills. Faced with so many problems, and seemingly unable to do anything which might ease the situation, Ingram's patience gave out. Dole was trouble enough, but all of a sudden it seemed to Ingram that Kennedy was also offering almost no assistance. In a weak moment, Ingram suggested that it did not seem unreasonable to expect that Kennedy should have succeeded in disposing of *some* lumber at *some* price and unfairly admonished him not to "hold it on speculation and ruin our credit by doing so."

More hurt than angry, Kennedy could only reply that, as they both knew, he had never considered holding the lumber on speculation. Indeed, he had made every reasonable effort to sell, again explaining: "It is not the price that hinders but the want of money and the large stocks already on hand." He then added, with mild indignation,[42]

> I think if you were to consider all things, the difficulties that I have had to contend with in getting the lumber out in the middle of winter, and the extreme tightness of the money market and consequent dullness of business of every kind, you would not think that I have done so very badly after all; there is a vast difference between getting out lumber in the summer season and now, a greater still in selling it: in the fall dealers buy all that they want and don't calculate on getting any more till Spring,

[41] Kennedy to Ingram, December 24, 1859.
[42] Kennedy to Ingram, January 24, 1860.

and it is only a chance to sell any during the winter, and especially such a winter as this has been. I don't say that I have done all that could be done, but I have done all that I could do, and I have got to see others placed in the same situation do better before I will think that I have not done well enough.

Any dissonance between Ingram and Kennedy disappeared in the face of more pressing considerations. By the first of March, Kennedy was back in Eau Claire doing the things he did best and most enjoyed. He and Ingram worked together, reviewing the plans and completing the details for the commencement of a new and hopefully more profitable year. Like the managers of a losing team, the mere thought of a new season and a new start revived their optimism.

But if the Eau Claire partners were able to forget the disappointments of 1859 as they looked ahead, Alexander Dole in Ottawa enjoyed no such relief. Part of the reason the absentee partner could not share in the excitement of spring and the expectancy of the coming season was that he was still awaiting official notice of just how poorly Dole, Ingram & Kennedy had fared during 1859. He waited none too patiently. Even if Ingram neglected to keep him informed by letter on any regular basis, Dole felt he surely had the right to expect prompt receipt of the annual business statement. But January of the new year came and went without a word, and the old man could only express amazement in mid-April that "those everlasting Books are not yet Ballanced. . . ."[43] Ingram was obviously stalling. He knew that Dole would be unhappy regardless, and Ingram did not care to review the past year just now. There was no time for critiques and criticism; only for planning ahead.

A good measure of Ingram's confidence in the spring of 1860 resulted from his success in arranging for sale of some lumber even before the sawing started. During the winter months he had been hard at work, corresponding with dealers in an attempt to contract as much as possible of the 1860 product of the Dole, Ingram & Kennedy mill. With the terms of sale prearranged, the more routine responsibilities of delivery, receipt, and payment could largely be left to agents. It was for this purpose that the services of Isaac Plume, soon to assume the task of keeping the

---

[43] Dole to Ingram and Kennedy, April 12, 1860.

books at the East St. Louis yard, and James Roberts, a political
activist in the new Republican party, were retained. For the
present, an old friend from Ottawa and early employee of the
firm, Ebenezer H. Playter, was left in the yard to be responsible
for selling whatever lumber might reach that point. It was
hoped that these arrangements would leave Kennedy free to supervise the operations in the mill while Ingram took charge of the
general management of the business.

All of Ingram's plans and preparations, however, could not
eliminate one large source of grief: Dole. In this instance, Dole
objected to the hiring of strangers and giving them responsibilities of a high order. He argued that even were Ingram successful
in contracting some of the lumber produced, it was unreasonable
to expect to arrange prior sale for all of it. Surely his partners
were not planning to entrust the balance of their lumber to just
anyone, for, the reluctant Dole maintained,

> . . . especially on the Mississippi I think it is necessary for sales
> to be looked after by some other person than a comparative
> stranger as there is circumstances attending the market there that
> requires consideration and changes to meet as they occur without
> having to discuss the matter with persons at a distance before
> Making sales, or being too much hampered by instructions or
> orders.

Accordingly, he advised that if Ingram had decided that it was
necessary for him to remain at the sawmill in Eau Claire, then
"Mr. Kennedy should look after the sales till I was able to go
out. . . ."[44]

The appearance of this objection proved to be but the introduction to a problem which would trouble Ingram time and
again in the years ahead. The cause of these growing complexi-
ties was obvious enough. In many areas of production, lumber
included, American industry was hesitantly entering a period
of its most rapid development. This entry was hesitant for
many of the nation's business firms because in order to expand
production and reduce per-unit costs a very difficult decision
was inevitably required: the old owner-operators had to give
way, or at least become partially dependent upon hired managers,

---

[44] Dole to Ingram and Kennedy, April 12, 1860.

the professional executives. But this was a slow and frequently painful transition.

Just as Alexander Dole objected in 1860 to the employment of "a comparative stranger" in a position of responsibility, so Kennedy and Ingram's subsequent partners would express misgivings for the balance of the generation. In the interim, changes in attitude did occur, but the fears remained much as before. For example, the common practice of the 1880–1890 period was to employ as supervisory personnel only men who held some amount of stock, thereby satisfying the traditional view that one manages well only that in which he has a personal interest. Managers under such arrangements were expected to live frugally on small salaries, content in the knowledge that their few shares of stock, which had generally been purchased by means of a loan from the major stockholders, would provide a bit of future wealth and security. Thus in 1890, when one of his valued managers sought an increase in salary, Ingram was probably not surprised by the reaction of a conservative associate who noted that he had seen[45]

> . . . a late letter of Mr. Ulrich's to my brother in regard to that matter, in which he still urges that his salary should be increased to $3,000.00 and intimates that if it is not increased to that amount, that he will have to look out for something in which he can do better. Now that expression I dislike very much, and if that is his feeling, I think that the stock holders or officers of Rice Lake Co. should be on the look out and see where they could get some one to fill the place. I have never thought it good policy to keep any stock holder in the employ in any department of our business where they were looking to the salary as the leading cause of their remaining in the company, and this request with almost threatening demand . . . is, to my mind outrageous.

The question of an increase in salary for Ulrich, the manager, remained a matter of contention for more than four years and was finally resolved by an indirect but apparently satisfactory maneuver. At the annual meeting of that company in June, 1894, the directors voted Ingram a $500 increase in salary with the tacit understanding that he would, in turn, give it to Ulrich. Deference was thereby accorded to the old beliefs and, at the same time, a

---

[45] D. M. Dulany, Sr., to Ingram, May 5, 1890.

professional executive was kept employed and reasonably content.[46]

But the Ingram of 1860 could not have foreseen that much of his future energy would be devoted to such problems of personnel. Actually, the management questions raised by Dole deserved and had received consideration. Prior to the completion of the rail network, especially the filling in of north-south routes, communication between Eau Claire and Mississippi River town was slow and uncertain. From their experience, all three partners realized that the sale of lumber rafts often depended upon the ability to reach an immediate decision as to price and terms. The expenses of running lumber could not tolerate frequent delays, and potential buyers could not be expected to wait for an agent to communicate with a distant owner.

In any event, when the first raft of the season left Reads Landing in April, 1860, Ingram was aboard. He was fully aware of the problem and in spite of the fears expressed by Dole, he was not one to shirk responsibilities. Although spending time downriver with the rafts was something Ingram had hoped to avoid, his efforts of the previous winter nevertheless gave some indication of success. Gradually, and at long last, conditions improved in the lumber markets of the west as the season wore on, and it began to appear that the worst of the depression had been weathered. Still, all problems did not disappear or even diminish in the glow of better times. In some respects, it became more obvious than before that the role of the East St. Louis yard within the Dole, Ingram & Kennedy organization would have to be clarified. Was it to become important in and of itself, as Dole foresaw; or was its purpose merely to serve as a catch basin for unsold lumber, as Ingram contended?

The improved conditions in the lumber markets would assure the ascendancy of the position maintained by Ingram. With the increased demand along the river, less and less lumber reached St. Louis, with the result that the assortment in their little yard soon became insufficient, at least from the standpoint of permitting equal competition with neighboring dealers. Ingram and Kennedy were entirely aware of the situation. Even their bookkeeper at the East St. Louis yard had reminded them that they

[46] Ingram to D. M. Dulany, Sr., January 22, 1895.

could hardly expect any profits from that point unless they were willing to maintain an adequate stock. Plume had complained in mid-August, "we have no dry stuff except some joist and scantling and are entirely out of what we need more than anything, viz., Fencing."[47]

But when business continued to improve, Plume had less difficulty accepting the fact that the yard would be largely ignored for the balance of the season, and for good reason. To Ingram he observed, "Your idea is no doubt to realize as soon and as much for your lumber as possible," which, he agreed, "is always desirable." As further indication that he understood, Plume added, "Lumber piled here . . . is money out at good interest."[48] Although they were denied the lumber needed to fill out their broken assortment, Plume and Playter nevertheless were able to advance yard prices, and before the end of September they reported selling no grade of lumber for less than thirteen dollars per thousand feet.

Partner Dole, however, was considerably less willing to accept the decisions made in Eau Claire than bookkeeper Plume. For example, when Ingram arranged a trade with Cruikshank & Son, a Hannibal wholesaler, Dole accused him of selling lumber for ten dollars a thousand which had originally been promised to the yard in East St. Louis where it would have brought a higher price. This, Dole charged, *"was anything but a judicious change."*[49] It was pointless for Ingram to explain once more that Cruikshank had offered cash, and that Dole, Ingram & Kennedy still needed cash more than they needed the promise of better prices at some indefinite future date.

As their second full year of operation drew to a close, Dole was sadly unable to share the confidence growing out of the general improvement in the market. Not even this first taste of success could restore his faith in Ingram's ability to manage. Indeed, as far as Dole was concerned, the problems seemed larger than before. Money he had previously borrowed was fast falling due, and although the company had marketed $25,000 worth of lumber during the 1860 season, somehow the Ottawa partner

[47] Plume to Dole, Ingram & Kennedy, August 11, 1860.
[48] Plume to Ingram, September 24, 1860.
[49] Dole to Ingram and Kennedy, December 2, 1860.

was like his pet project, the East St. Louis yard—they both got what was left over, which did not amount to much at either place.

However unhappy Dole may have been, the firm of Dole, Ingram & Kennedy was established. The partners had constructed a good sawmill on an ideal location. In the process, they had increased their expertise as lumbermen. Procurement of the pine resource was no problem, and they had recently begun to experience success in marketing their product. Ingram and Kennedy were aware that they had come a long way in a relatively short time.

Although much had been achieved, two major areas of difficulty loomed large. The first involved the continuing deterioration of relations between Dole and Ingram, with the latter already looking forward to the day when he might be able to buy out the interest of the older partner. The second area of difficulty was the continuing deterioration of relations between North and South. Should political conditions stabilize, further and more rapid progress in coming seasons might reasonably be expected for the young firm. But for Americans in 1861, political stability seemed a most unlikely possibility.

CHAPTER IV

Wartime Prosperity
and New Vistas

POLITICAL CONSIDERATIONS ASIDE, Dole, Ingram, and Kennedy viewed the coming of the Civil War as a most unfortunate and unwelcome development, bringing new concerns and uncertainties at a time when the firm seemed on the verge of profitable operation. That the war might also prove a boon to the lumber industry was no solace. Indeed, the partners would gladly have exchanged such a possibility for the assurance of an even chance to sell their lumber in a relatively stable market.

This, however, was not to be. The markets of the wartime economy were subject to dramatic fluctuations which for the most part greatly benefited the manufacturers and producers of the North. While the lumber industry did not prosper as directly and as quickly as iron, leather, and the like, or even agriculture, its profits were nonetheless substantial. This was particularly true beginning in 1863, when the demand for lumber sharply increased as farmers, having already settled their old debts, directed surplus funds towards new purchases and improvements. That much of the currency recently issued was commonly regarded with suspicion served further to encourage investments of this sort. In many respects, Dole, Ingram & Kennedy were representative of the small lumbering organization of the day. After an initial period of uncertainty which included the loss of their sawmill by fire, they soon began to reap the benefits of better times. By 1863 they were enjoying good times. When the wartime prosperity ended, the adjustments which followed were withstood with relative ease by a company that was far more substantial than it had been five years earlier. Soon, expansion and emigra-

69

tion and all that was associated with continuing national growth more than took up the slack.

But for Dole, Ingram & Kennedy, the 1860's brought considerably more than big business at high prices. Not unexpectedly, the partnership was reduced to two when Dole finally concluded that he had had enough. The promise of an eventual return on his Eau Claire investment was no longer worth the mental anguish that he was forced to endure in the process. When he began to realize all that was involved in an independent operation, he began to appreciate his own limitations. He thought he had wanted to start over in the pineries of the Chippewa, but in truth he was too old and too tired to make the effort. Dole had been of inestimable importance to Ingram, and the younger partner would make the older man's dreams a reality. To do so, however, Ingram needed only Dole's money, not his advice. From the beginning, the company had really belonged to Ingram. Accordingly, Dole's official withdrawal early in 1862 had little discernible influence on the manner in which business was done. Fortunately the new partnership of Ingram & Kennedy had also progressed to the point where Dole's sources of eastern credit were of less critical importance than previously.

There were changes more significant than the departure of Alexander Dole which occurred during the decade. Many of these resulted from an increase in mill capacity with the accompanying requirement for more sawlogs and for larger areas for log storage, as well as the necessity to give greater attention to marketing. These efforts to expand production were common to lumbermen throughout the area. In the process, the value of pine was enhanced where it grew in accessible stands. Thus the forests of the Chippewa valley assumed an almost overnight importance, not only to provide for the needs of the local operators but also for competing millmen of other valleys who had no dependable supply of their own. By the late 1860's it was apparent to a reluctant Ingram that he would soon be forced into active competition for a future supply of pine, though at that time he could not foresee the extent and expense of the conflict which would grow out of that competition. In the early years of the decade, Ingram was far more concerned with the tragedies of the war.

Whatever its principal causes may have been, little in America

was untouched or unchanged by the Civil War. Still, there
were times when the issues and the drama of the war seem to have
been neglected if not ignored by contemporaries—especially if one
limits his purview to the records and correspondence of Northern
companies which were straining to take full advantage of un-
imagined opportunities. But such an impression is not altogether
justified. While businessmen may not have allowed their earlier
opposition to interfere with opportunities once the battles had
begun, this was one war which demanded considerably more than
mere profit-taking, even for the most venal speculators. Pas-
sions ran too high and the issues were too emotional to permit
detachment.

This did not mean that businessmen as a body marched off to
do battle for the Lord. Regardless of how strongly they felt for
the righteousness of their cause, well-to-do members of northern
communities normally allowed the patriotism of their less affluent friends and neighbors to fill the ranks, and when patriotism
alone no longer sufficed, bounties usually made up the draft
quotas. Like most of their peers, Ingram and Kennedy remained
at home on the job.

In the case of Ingram, moreover—thirty-one years old and in excellent health—his dedication to business was reinforced by his
political attitudes. While not a "Copperhead," he was a loyal
Democrat to whom the prospect of Abe Lincoln as President was
considerably less than cheering. For a rather brief period during
the winter of 1860–1861, Ingram welcomed opportunities to discuss the issues of the day, expressing his belief that anyone who
was truly interested in peace should support the cause of patient
compromise. But these were clearly minority views along the
banks of the Chippewa, and even had Ingram's forensic skills been
of a higher order, it is doubtful that his reasoning would have
gained much acceptance. More likely, the participants in such
discussions agreed with Dole, who maintained that *"Right* is just
and must prevail," and who wrote Ingram from Ottawa in December of 1860: "I think your theory on what is passing is as
much in error as it was in the vast majority you counted up for
Stephen A. Douglas."[1]

As conditions worsened and the arguments grew louder, In-

---

[1] Dole to Ingram, December 2, 1860.

gram withdrew to the less controversial confines of office and mill. Insofar as possible, politics could not be allowed to interfere with business. This lesson, so obvious in the climate of unbending opinion in 1861, was never forgotten. Although he retained a lively interest in politics throughout his life and lost box after box of cigars betting on an almost unbroken string of Democratic defeats at the local, state, and national level, his involvement remained largely private. He never had a doubt that the concerns of business came first. On those occasions when he ceased to be a spectator and briefly assumed an active role in politics, there was always a business consideration at the basis of that involvement.

The secession crisis and the threat of war summarily halted the improvement in conditions in the western lumber markets which had taken place during the late summer and fall of 1860. Throughout the winter months, when Ingram sought without success cash advances on lumber to be delivered in the course of the 1861 season, it became apparent that hard times remained or had returned.

The practice of wholesale dealers advancing money in the winter on lumber which they would receive the following spring and summer was very common. Cash paid to the millmen in these exchanges would likely be used to meet the expenses of logging. In fact, this practice became a basic part of an eternal debt-incurring process. Producers borrowed money in advance from lumber buyers for the purpose of procuring a supply of logs and to repair and improve their sawmills. Consequently the lumber, when delivered, having already been paid for in part on terms favorable to the buyer, brought less cash to the producer; and he would be forced to borrow again in order to meet the expenses of a new supply of logs and additional repairs on the mill. There were, of course, many other variables in a year's business transactions, but this was invariably the skeleton on which they were arranged.

In any event, it was not unfamiliarity with the procedure which prevented the advancement of funds on lumber during the winter of 1860–1861. Rather it was the uncertainties resulting from the deteriorating political situation. One of the firms on which Ingram had thought he could depend for an advance was R. Musser & Company of Muscatine, Iowa. But Musser complained in

February that "not a dollar can be had"; and while they would certainly differ as to probable solutions, Ingram could fully agree with the Iowa lumberman's assertion that "Political Demagogues will have an awful account to settle some day for the trouble they are making."[2]

Although unsuccessful in obtaining much in the way of advances, funds from an unexpected source provided brief relief to Dole, Ingram & Kennedy early that spring. James Roberts was selling some lumber from the little yard in East St. Louis. Business there had begun to improve in February, and during the first half of March Roberts took in nearly $700 in cash, which he personally viewed as quite an accomplishment considering the limited stock he had on hand. Better times seemed just around the corner and Republican Roberts predicted, no doubt partially for Ingram's benefit, "things will be better when old Abe's arrangements are completed." As for himself, in recognition of his reputation and the attitudes of his employers, Roberts reported: "I have been very good . . . not having taken a glass of beer for a month. . . ."[3]

But neither abstinence nor old Abe could prevent the rapid deterioration of business conditions in and around St. Louis. Time proved that Roberts' early spring customers were buying before the storm, and by mid-May the flurry of sales had ceased. W. H. Padon, a dealer in nearby Lebanon, Illinois, noted that all business activity had been suspended, explaining, "The only thing that occupies the minds of the People is preparations for war."[4] The bitter struggle for Missouri was about to begin. For the duration at least, as Padon advised, building and improvements of every kind would be forgotten in the St. Louis area, and "there will consequently be but little demand for lumber." The correctness of this prediction was shortly borne out, and by June there was almost no call for lumber in the St. Louis market. Musser even reported some expensive attempts at shipping the unwanted lumber from St. Louis back upstream to Muscatine. This, he observed, must be somewhat like "carrying coal to New Castle."[5]

[2] R. Musser & Company to Dole, Ingram & Kennedy, February 11, 1861.
[3] Roberts to Dole, Ingram & Kennedy, March 15, 1861.
[4] W. H. Padon to Dole, Ingram & Kennedy, May 14, 1861.
[5] R. Musser to Dole, Ingram & Kennedy, June 10, 1861.

In Eau Claire, however, sawmilling operations proceeded in an almost normal manner. There was little reason and less opportunity to do otherwise, although the demand for lumber had never been poorer than during the spring and summer of 1861. Alex Kempt, sent to Eau Claire from Ottawa by Dole in the hope that he might be able to unravel the puzzle of the firm's books, soon found himself a long way from Eau Claire and from bookkeeping. He celebrated the Fourth of July "waiting for the Raft" at LeClaire, Iowa, trying to make up his mind whether or not to let Musser have it for $7.50 per thousand feet. Since there seemed little prospect of a better price elsewhere, Kempt finally decided to accept the offer, fully realizing that the details of the trade were not likely to impress his employers. Therefore, when he reported the sale to the Eau Claire office, he carefully described his efforts to sell the raft to dealers in every market town along the river, adding, "they actually don't want lumber at any price."[6]

Conditions in the Mississippi River markets remained extremely depressed throughout the summer of 1861, and the piles of unpaid bills in Eau Claire reflected the downriver situation. Demand for lumber nearly disappeared. Lumbermen, however, were not alone in their difficulty, for there was little business activity of any sort. Iowa farmers could realize only thirty-five cents per bushel for wheat and ten cents for corn, neither price being sufficient incentive to go to the bother of selling. But if the farmers of Iowa complained of grain prices, they were at least better off than their brethren in Missouri and Kansas, who were fast realizing the true tragedy of civil war. For all practical purposes, business activity in St. Louis altogether ceased. Even Dole was compelled to agree that no more lumber should be permitted to go through to the yard at East St. Louis and that every effort should be made to sell what remained there at the earliest opportunity.

With such a deficiency of buyers in the Mississippi River markets, the partners had no alternative to the establishment of a second yard in which to hold their growing assortment of unwanted lumber. Thus, in the heat of early August, Eb Playter supervised the drawing, washing, and piling of eighteen cribs at Wabasha, Minnesota, a short distance below and opposite the

---

[6] Kempt to Dole, Ingram & Kennedy, July 4, 1861.

mouth of the Chippewa.[7] Later in the month, after Playter had returned to St. Louis, William V. Gill assumed management of the new yard.

In their efforts to obtain money from some source that summer, Ingram concluded to pile a small amount of lumber at the mill for possible sale to Eau Claire customers. Despite the large amount of lumber produced in the vicinity, perhaps as much as 50 million feet in 1861, little attention had been given the local market. Accordingly, the endeavor was noted in the local newspaper, the editor observing that "Nothing ought to be more gratifying to mechanics and builders, than to know that seasoned lumber can be had."[8] Not that Ingram was particularly solicitous of local mechanics and builders; he was simply willing to try almost anything which might result in cash sales.

By the fall of 1861, the worst of the business crisis was over. The need for lumber clearly could not await the end of war, and buyers with cash began to examine long-ignored piles of boards with new interest. Playter reported from Muscatine in late October that demand was much better than the dealers there had foreseen, Musser expressing regret that he had not "given Kempt an order for another Raft when he left here, and as he heared you were piling lumber at Wabasha, thought you would not be able to supply him, so he bought of Washburn and Knapp, Stout & Company."[9] But if they missed out on a sale to Musser, they missed few other opportunities. Business looked even more promising than it had a year previous. Lumber was sold, bills were collected and paid, and plans for the future proceeded with renewed confidence.

This satisfaction of making profits and plans suddenly ended. On the 26th of October, in the darkness of an early Sunday

[7] Lumber that was moved by water naturally required some attention after arriving at its destination. By various means the cribs—those tighlty bound, squared stacks of boards and timbers which made up the raft—were removed, or drawn, from the river to the bank. In the course of the journey downstream, the lumber would collect sand and silt, so the boards were literally washed off before being taken to the yard. There the lumber was piled in a manner which would encourage drying, a process which would hopefully be completed prior to marketing needs. As the years passed, planing mills were established at most downriver wholesale yards by which means the boards, stained and dirty and even warped as a result of their watery transit, could be made bright and clean again.

[8] *Eau Claire Free Press,* June 14, 1861.

[9] Playter to Dole, Ingram & Kennedy, October 22, 1861.

morning, the Dole, Ingram & Kennedy sawmill caught fire and burned to the ground. Since this was the initial encounter with such adversity for the lumbermen of Eau Claire, and indeed one of the town's first serious fires of any sort, no organization had yet been established for the purpose of fighting fires.[10] Whatever its cause, the fire spread so fast that the night watchman's efforts and alarm were to no avail. Nothing survived except a little of the machinery which resisted the flames and intense heat. The loss was estimated at $15,000.[11] The circumstances of this unfortunate conflagration were never forgotten. If in later years partners and employees grew weary of Ingram's admonitions about the hazards of fire, it was in part because they had not shared this early experience.[12] A preoccupation with the constant dangers of fire was mandatory for the successful lumberman.

Yet even as they totalled up their losses in the fall of 1861, the partners must have realized that they had good reason to be thankful. The destruction of their mill coincided with the end of the sawing season. They had already used up all of their logs and sawed out all of the lumber that would have been produced until the beginning of the next season, fire or no fire. Of equal importance, should they determine to rebuild the mill, there was time enough so that at least a portion of it could be "in running order 'as soon as the logs come down' in the Spring." Had the fire occurred six months or even three months earlier, the partners might well have been forced to give up the effort at Eau Claire. As it was, within two weeks of the fire they announced plans to

[10] It seems incredibly fortunate, if true, but "the first fire of any account that has ever happened in our town" was reported in the *Eau Claire Free Press* of January 24, 1861. Not until November 15, 1869, did the Eau Claire trustees take any action in the way of providing municipal fire protection. On that evening they approved the purchase of "hooks and ladders, and the erecting of a small building in which to store them in order that they may always be found at the proper place . . . in a moment's notice." *Eau Claire Free Press*, November 18, 1869.

[11] *Eau Claire Free Press*, October 31, 1861.

[12] In mid-summer, 1892, Ingram offered what had become a very common exhortation. Writing to one of his partner-managers, he expressed the hope "that all of you will be on guard all the time against fire." Continuing, Ingram admitted, "I worry more about fires in the different places where I am interested than anything else, and I talk about it so much that parties some times I think call me a crank on that subject." But such a possibility did not prevent him from making the point again. One could not be too careful. The new sawmill at Hudson, Wisconsin, would "burn up in about 15 minutes, if it got a good chance." Ingram to George P. DeLong, July 5, 1892.

rebuild. There was no question but that Dole, Ingram & Kennedy would "soon have another, and if possible, more perfect and complete mill than before."[13]

Ingram and Kennedy decided to rebuild their sawmill long before all arrangements for funding had been completed, reflecting confidence in the future rather than the certainty of current resources. They were encouraged to no small extent by the recent improvement in the lumber markets downriver. Unlike the previous year, when Ingram had been unsuccessful in arranging advances on their lumber, he now obtained, with some concession in price, advances on lumber already sawed to be delivered as soon as the ice went out as well as on the lumber which would be sawed in the new mill during the coming season. That the profits would be smaller on lumber thus contracted was of less importance than the fact that they were able, because of their reputation, to arrange such contracts. Again, the critical problem was to obtain cash. It was hardly a time to be much concerned about margins.

Their success in obtaining money on future deliveries of lumber was not simply the result of improved economic conditions. In the misfortune of losing their sawmill, the partners happily learned that they now had some friends and business acquaintances willing to provide assistance. There were others who hoped for the success of Dole, Ingram & Kennedy, partly because of the quality of their lumber and partly because of a special regard and concern for the owners as individuals. It therefore seems likely that considerable help came from dealers in and around St. Louis. Whatever motivations might have influenced a decision to invest in the rebuilding of an Eau Claire sawmill, there was almost no use for surplus funds in St. Louis during the winter of 1861–1862.[14]

[13] *Eau Claire Free Press*, November 14, 1861.

[14] It is not clear from the extant records how they arranged for the capital necessary to finance the reconstruction, and it is especially interesting that they managed to do so in such a short time. There had been no insurance on the old sawmill. Also, considering the strained relations then existing, it was unlikely that Dole would have been willing to advance any more personal funds or even agree to seek much in the way of additional credit from Canadian or New York sources. This is not to say that the Ottawa partner attempted to discourage the plans for rebuilding. He doubtless favored such an effort as the only possible way of getting back what he had already invested, for by this time Dole surely had decided to sell out his interest.

By various means and from a variety of sources, the necessary monies were secured and plans for the new mill soon began to take shape. A few weeks earlier Playter had finally succeeded in negotiating a trade with W. H. Padon for the balance of lumber left in the East St. Louis yard. Playter was pleased and relieved for a number of reasons. The arrangements he had made for the lumber were as good as could be expected under the circumstances, with the added assurance that "whatever Mr. Padon promises to do, he always does."[15] Also, Playter was anxious to leave St. Louis, which he found dirty, depressing, boring, and distant from his bride-to-be, an Eau Claire girl.

The season of 1862 proved better for business than any previous period experienced by Dole, Ingram & Kennedy. The new mill was ready for operation in the spring, according to plan. By late March, following the Union victory at the battle of Pea Ridge in Arkansas, Padon reported slight improvement even in conditions at St. Louis, the most important being that "I am getting the money for it when I sell."[16]

With the approach of the rafting season, orders for lumber began to arrive at the Eau Claire office with a frequency not previously experienced. New customers inquired and customers long silent wrote again. The little Andulusia, Iowa, dealership of Buffum & Brothers noted that they had been out of touch with Dole, Ingram & Kennedy for so long a time that "we hav come to the conclusion that you ware Broke, Mill blown up or Some other awful thing had taken Place." Still, they ordered some lumber on faith, observing that "there is the best Show for our business this Spring there has been for the last 3 years." Their fencing was selling at twelve dollars per thousand feet and was "very Scarce."[17]

The improvement in the market, however, in no way lessened the determination of Alexander Dole to leave the company. In fact, the time of final settlement may have been advanced by the increase in demand and prices, which enabled and encouraged Ingram and Kennedy to assume the responsibilities of payment required to purchase Dole's interest. Dole remained resolute in his decision to withdraw, and the Eau Claire partners, who had

[15] Playter to Dole, November 2, 1861.
[16] W. H. Padon to Dole, Ingram & Kennedy, March 27, 1862.
[17] Buffum & Brothers to Dole, Ingram & Kennedy, March 18, 1862.

long been willing, were at last able to comply. Before the fire their creditors had been many. Construction of the new mill increased their indebtedness by a large amount. The terms of payment to Dole left Ingram and Kennedy with still greater liabilities.

Although more lumber was sold and more money was made during the 1862 season than in any previous season at Eau Claire, there were more than enough bills to absorb the increasing profits. Cooley, Farwell & Company, a Chicago wholesale drygoods firm which included young Marshall Field, was still demanding the balance due on a note dated June 1, 1861, announcing, in October, 1862, in terms unmistakably clear: "Grace has run out."[18] Marshall Field and his associates were far more likely to receive some portion of their due than were a few needy ex-employees, who were also awaiting settlement. One of these, a man by the name of John Patterson, included his request in a long and newsy letter, which he closed in the following manner: "Give my best respects to all the folks . . . as many as you think fit. That is all at present. But the money is the main thing."[19]

Nevertheless, the eternal problem of paying bills was slowly becoming less a burden, and this gradual easement was especially apparent in the case of long-term notes. When the Dole, Ingram & Kennedy mill was destroyed by fire, they were fortunate to incur their large indebtedness under favorable circumstances. Wartime inflation was beginning to affect the lumber market, with accompanying advantages to those in debt.

The payment of bills, of course, did not proceed speedily enough to satisfy all of their creditors, who unhappily watched the value of money owed them diminish as the inflationary spiral continued. One of these creditors, ex-partner Dole, lamented what seemed to him an unnecessary delay in forwarding the payments due. If he was not vindictive, he was not particularly solicitous of the many problems then encountered by Ingram & Kennedy. In September, 1863, he urged Ingram to make a final settlement, "without loss of time and delay in carrying out the transaction."[20]

[18] Cooley, Farwell & Company to Dole, Ingram & Kennedy, October 30, 1862.
[19] John Patterson to Ingram, October 3, 1862.
[20] Dole to Ingram, September 28, 1863.

Ingram naturally regarded such demands for haste as unfriendly and unreasonable. In his reply, he expressed surprise that Dole would consider calling upon them for a "payment in October after making so liberal a payment in July, nearly one half of the whole indebtedness," adding his own opinion, "We feel that we have done well by you." Furthermore, Ingram pointed out that he and Kennedy were at least as anxious as Dole to wind the business up "and have the Interest stopped." In conclusion, Ingram reminded his old nemesis that there remained "some other indebtedness of the Firm of Dole, Ingram & Kennedy that we have to provide for, or else you as well as we would have trouble."[21]

In his letter to Dole, Ingram attempted to describe the nature of some of the recent difficulties they had encountered at Eau Claire. The most serious of these resulted from too little water in the Chippewa, allowing them to prove once again the truth of the lumbering adage, "low water makes rusty saws." They had failed to get an adequate supply of sawlogs to their mill and were therefore unable to make the amount of lumber which they had planned. Despite these and other problems, however, Ingram assured Dole that it seemed "quite certain" that they would be able to make him a payment sometime in December. For all of Dole's complaints, his former partners clearly had made substantial progress since his departure from the firm. Indeed, it seems likely that it was the success of Ingram & Kennedy rather than the schedule for payments which most annoyed Alexander Dole.

Obviously, this success was not all of their own doing. It may have seemed bad form to thank a war, especially a civil war, but the benefits to business and industry came nonetheless. As the violence raged into its third year, prices continued to rise and the lumber markets continued to improve. In January, 1864, normally a month of small sales, Gill reported the largest trade yet experienced in a single week at the yard in Wabasha, amounting to more than $1300, or an average of $225 for each working day. So much money on hand made Gill uneasy, and he asked what arrangements had been made for sending cash to Eau Claire.[22] Ingram must have welcomed the opportunity to consider such

[21] Ingram to Dole, October 6, 1863.
[22] W. V. Gill to Ingram & Kennedy, January 18, 1864.

problems. During the month of January, Gill sold more than 110,000 feet of lumber, and the Wabasha yard, begun for lack of an alternative, began to look like a wise investment.

But record sales at Wabasha were only one indicator of rapidly improving conditions, and Ingram & Kennedy was not the only firm on the river enjoying the prosperity. Chapman & Thorp, selling more than they could supply with their own lumber, offered to buy a hundred thousand feet from Gill in Wabasha. In previous seasons such a sale would have been considered a splendid opportunity, but better times had lessened the need for cash at any cost and Gill was in no hurry to close the deal. Instead he sought Ingram's advice, noting that it really mattered not at all whether he sold the large amount to Chapman & Thorp, since he would surely sell it to someone else by the first of July.[23]

The relative proximity of Wabasha to Eau Claire held a good many advantages, the most obvious of which was the cost of running lumber. Because Wabasha was so close to the mouth of the Chippewa, rafts could be run there directly out of the tributary with no intermediate arrangements necessary for the purpose of coupling the cribs to make the larger Mississippi River rafts. This not only reduced expenses as a result of the time and distance involved, but also eliminated all of the costs associated with coupling, watching, and holding the lumber rafts at Reads Landing. In addition, the arrival of lumber tended to be on a more regular basis than was the case at points farther downstream, and the Chippewa raft pilots could act as couriers, reporting current yard needs back to the sawmill and carrying the cash from the Wabasha yard to the Eau Claire office. In successive weeks in late April and early May, 1864, for example, Gill forwarded bundles of cash containing $800, $600, and $500 by pilot William Miller.[24]

In spite of all of these advantages resulting from proximity of yard to mill, the supply of lumber could not keep pace with the demand at Wabasha in the spring of 1864. Gill's report to Eau Claire in late May clearly indicated that selling was not quite the same chore it had recently been. He told of the Chippewa pilot, Miller, arriving with a raft the evening previous; no sooner had

[23] Gill to Ingram, April 5, 1864.
[24] Gill to Ingram & Kennedy, April 30, May 7, and May 13, 1864.

the raft touched shore than the six cribs were on the bank and all the fencing, some 5,500 feet, was sold. The impatient buyers "took part of the Lbr. out of [the] water themselves They were so anxious to get the fencing." That evening and the following morning, Gill loaded twenty teams with lumber out of the little yard in Wabasha. Indeed, times were good and getting better. At the end of each working day Gill had less lumber and, on the average, about $200 more cash.[25]

During May, nearly 200,000 feet of lumber were sold at Wabasha, along with 72,730 shingles and 41,500 pieces of lath, and Gill complained that he could easily have sold another fifty thousand feet "if I had had it."[26] With demand already exceeding supply in the yard at Wabasha, he could not but resent any apparent diversion of lumber elsewhere. But one of the benefits resulting from just such a yard was the flexibility it offered in the marketing operations, particularly during hard times. By neglecting the downriver markets in order to keep Wabasha well-supplied, they would, in the process, lose their access to those markets. Ingram realized, if Gill did not, that while things were anything but dull at the Wabasha yard in the spring of 1864, there would be other seasons and other circumstances.

By the middle of June, prices for common lumber at Wabasha had increased to twenty-two dollars per thousand feet, and Gill was sending cash in thousand-dollar packages to Eau Claire. Conditions for the sale of lumber could hardly have been better—providing, of course, that there was lumber to sell. Gill advised that only about forty thousand feet remained in the yard. Obviously some change in policy had to be considered, and even Ingram proved unable to resist the promise of such profits. Not only did he commit the firm to keep Gill in good supply for the balance of the season, but he and Kennedy also decided on the basis of their successful experience at Wabasha to open another little retail yard at nearby Minneiska, Minnesota, hiring J. B. ("Jabe") Randall of Eau Claire as manager.

Although Randall welcomed the opportunity of selling lumber for Ingram & Kennedy, observing that the residents in and around Minneiska offered quick assurance that the yard would

[25] Gill to Ingram & Kennedy, May 25, 1864.
[26] Gill to Ingram & Kennedy, June 13, 1864.

have all the business he could possibly manage, he was only cautiously optimistic as he completed his preparations in early August. He agreed to "give it a Tryal"; at the moment, however, he seemed far more concerned with Eau Claire's progress in "Filling our Quota of Volunteers. . . ."[27] The battlefields were distant but the war continued. Grant and Lee faced each other before Petersburg, Atlanta still held out against Sherman, and the casualty lists grew ever longer.

If Ingram had experienced some occasional difficulties in satisfying all branches of the company before Randall settled at Minneiska, that experience had been only an introduction to what would follow. Jabe Randall recognized immediately that if he were going to make a name for himself at Minneiska, it would be in spite of Mr. Gill at Wabasha. The two yards were too close not to be competitive, even though Gill had been assigned a general responsibility for the management of both. Randall's understanding of the competitive aspect of the situation was evident in his first letter back to the Eau Claire office. With no acknowledgement of subordination, he requested simply and directly that he be given "as good a show as [Gill] so he will not draw Trade from here."[28]

Specifically, Randall referred to keeping his own yard sufficiently supplied with lumber so that he would be assured a fair chance to sell to customers who might otherwise purchase from Gill at neighboring Wabasha. But Randall was also describing, in plain terms, a larger and constant theme of some interest and importance. It was not pettiness alone which caused him to suggest that while both yards were working towards the common objective of making a profit for Ingram & Kennedy, they were nevertheless in real competition with each other. Indeed, Randall realized that whatever the total sales might be in southeastern Minnesota, the Eau Claire office would judge his specific contribution solely in terms of the return from the investment in Minneiska. Although it now seems remarkably unsophisticated, Ingram & Kennedy, and doubtless many other firms as well, tended to judge the relative worth of their various branches on a strict profit-and-loss basis. The guideline was simply that

---

[27] Randall to Ingram & Kennedy, August 4, 1864.
[28] *Idem.*

each separate unit must prove itself by paying for itself. Success was then measured in terms of relative profits.

The yards at Wabasha and Minneiska presented the first occasion for such comparison within the Ingram & Kennedy organization. Whatever the inaccuracies and inadequacies of so simple a formula, it proved generally correct over the long term and in the majority of instances. Still, it is possible to imagine all manner of situations in which a single yard might of itself not be profitable but might serve considerable advantage to the larger organization. This would be particularly true when considering the opportunities for monopoly. But lacking a bevy of marketing experts, the rule remained in effect at least until small losses were no longer intolerable, and that was a long time afterward.

Business at the Minneiska yard began well, and Randall grew more optimistic as he had opportunity to see the land adjacent to his new home, with crops "never Better or Larger." As further incentive to the sale of lumber, in addition to the good times and the promise of a bountiful harvest, Randall reported the people to be "afrade of the money" and he felt that farmers would soon begin to "fence & Build some considerable" rather than hold on to their greenbacks.[29] Within a month after opening the yard, Randall informed the Eau Claire office that he was getting "vary anxious" for more lumber, especially wide boards, and that his shingles and lath were also in short supply. Never forgetting his competition with Gill at Wabasha, Jabe also took occasion to note that at Minneiska wheat was averaging $1.60 a bushel, "While at Wabasha it has been $1.52 so you see Wheare you will sell your Lbr. . . ."[30]

There were times when natural conditions frustrated all efforts to take advantage of big demand at high prices. That was the situation in the middle of September, when Randall began to experience the same difficulty that had troubled Gill earlier in the summer: the impossibility of receiving lumber as a result of low water. Oftentimes rafts would reach the mouth of the Chippewa before grounding on a sand bar or snag, sometimes within sight of the Minnesota shore. Waiting impatiently for the water to

[29] Randall to Ingram & Kennedy, August 13, 1864.
[30] Randall to Ingram & Kennedy, September 3, 1864.

rise and the rafts to arrive, Randall informed his employers that he could be selling ten times the lumber he had sold if there had been any decent assortment in the yard, noting that his fencing was all gone, so too were the wide boards, the best grade of flooring, and most of the shingles. Still, the customers continued to arrive and purchase lumber of some sort or another, and they continued to pay big prices. Randall was selling common lumber for $28 a thousand, and inquired of his employers, only half in jest, "is that High Enough?" Even the little yard at Minneiska returned cash to Eau Claire by the thousand-dollar bundle in the fall of 1864.[31]

As the price on lumber spiraled upward, single boards assumed an added value, so that for a time petty thievery became a problem at the yards. Randall suggested that a fence be installed around the perimeter of the Minneiska yard. In an effort to apprise the Eau Claire office of his concern about matters of security, Jabe described an attempt by a woman to steal a board in the light of mid-day. Although he chased her, caught her, and "made her bring it Bac[k]," he readily admitted that he had not relished the task and that for a member of the fairer sex, she was "Prety Scearry."[32] Because such incidents provided good story-telling material, the problems resulting from petty thievery probably received more attention than was their due. While Ingram and Kennedy and their associates were certainly not in business to provide free boards to the daring needy, losses from such causes were of minuscule importance, especially during seasons like 1864.

A larger but scarcely more genuine concern involved the apparent high wages paid to the laboring force in all phases of the lumbering operation. Wage increases were in part, of course, the result of inflationary pressures. For all of management's talk about the high cost of labor, however, real wages actually lagged behind price increases. Any employer who maintained a reasonably accurate set of books realized this, although he admitted it rarely and only in select company.

But employers did have reason to worry over the number of workers available, for there seemed to be no end to the number

---

[31] Randall to Ingram & Kennedy, September 16, 1864.
[32] Randall to Ingram & Kennedy, October 2, 1864.

of men required to fill the ranks of the Union armies. In this connection, the most direct contribution made to the war effort by the lumber industry would seem to have been that so many of its workers became soldiers. Thus it was with rather mixed feelings that Ingram and Kennedy watched such stirring events as the departure for Camp Randall of Wisconsin's 8th Infantry Regiment—the Eau Claire Rangers—who would soon be better known as the Eau Claire Eagles, in honor of their famous feathered mascot, Old Abe. If thoughts of the men, their families, and their sacrifices occurred to Ingram and Kennedy as they listened to the martial music and speeches, the partners' reflections were intermixed with concerns over how they would be able to replenish their dwindling labor force at the sawmill and what wages would have to be paid those remaining at home. Regardless of how high the price of lumber might rise, wages were clearly another matter.

The combination of a shortage of workers and inflation made its effects increasingly evident as the war continued, especially at those times of year when mills and yards were forced to compete for workers with farmers. In the harvest days of the summer of 1864, Randall complained that he was unable to hire yard help at Minneiska for less than four dollars a day, since men could earn that much and more by working in the fields of his customers. When an early winter followed on the heels of President Lincoln's re-election, the Minneiska manager continued to fret over the difficulty "to get help since snow come." According to Randall, the labor situation had deteriorated to a level where, in order to get anybody to work, you had to pay them high wages, "give them plenty of treats, and then do as they say."[33]

Despite the tenor of complaints at the owner-management level, workers were probably not as well off as they had been a few years earlier. As a case in point, if wages had doubled in the course of the war, lumber prices had trebled, and lumber prices merely reflected the general business trend. If costs seemed high in an absolute sense, Ingram & Kennedy was nonetheless operating at a considerable margin by 1864, allowing satisfaction of indebtedness at a much faster rate than even the optimistic Ingram could have foreseen.

[33] Randall to Ingram & Kennedy, November 16, 1864.

Business activity in the upper Mississippi Valley quieted down with the arrival of cold weather and heavy snow in mid-November of 1864. The pause gave the first opportunity for reflection for some time and the lumbermen began to wonder what the next season might hold. It seemed too much to expect that such ideal marketing conditions could last. Mr. Musser of Muscatine expressed doubt that farmers would continue to pay so dearly for lumber, noting that in southeastern Iowa they seemed more interesting in investing in land and delaying their improvements until the materials could be purchased at less expense.[34]

The big news from Wabasha in January of the new year concerned efforts to raise money to pay volunteers to fill their quota, Gill noting without objection that it would cost him about a hundred dollars.[35] Sherman had reached Savannah, but the war was not yet over. Jabe Randall in Minneiska was also concerned with the bounty, but more in terms of its effect on business, explaining the recent decline in sales by the fact that all extra cash was "going to pay the Volunteers." But he foresaw a return to busy days in the yard "very soon after the Draft takes Place."[36] It was not simply the direction of cash to bounty funds that slowed sales of lumber in early 1865. Sharing the responsibility to some extent were lower prices for wheat. Farmers, who had recently done much better, were naturally reluctant to market their wheat at a dollar per bushel, at least until circumstances forced them to do so. Although they might want lumber for improvements, they had to need it badly before they would sacrifice their grain to obtain cash. Unfortunately for themselves— and the lumbermen as well—their judgment proved faulty. In a short time a dollar per bushel would be a very high price for wheat on the frontier.

But if lumber sales seemed to hesitate at the two Minnesota yards, reports from St. Louis were encouraging for the first time in several seasons. Branch, Crooks & Company of that city, from whom Ingram purchased much of his sawmilling machinery for many years, thought the prospects were extremely good, noting that the St. Louis yards had almost no lumber on hand and that

---

[34] R. Musser to Ingram & Kennedy, December 29, 1864.
[35] Gill to Ingram & Kennedy, January 16, 1865.
[36] Randall to Ingram & Kennedy, January 18, 1865.

prices were very high. They advised Ingram to resume rafting to their city, confidently predicting "a first rate demand for all you can send down."[37]

While it was true that St. Louis had been starved for lumber during recent years, such was not the case in the river towns above. Indeed, the return of demand in the St. Louis market was one of the few encouraging signs evident to lumbermen as they looked ahead to the commencement of the 1865 season. Coincident with the certainty that Southern armies could not long withstand the mounting pressure of the Union forces, wholesale prices in general had begun to weaken. Although none could then foresee the extent or the long duration of the decline, few could doubt that the immediate future held the prospect of a deflationary adjustment.

Ingram accepted such postwar changes as inevitable. Unlike many of his acquaintances and a few of his associates, he saw no need for adopting the alarmist policies so commonly counseled. He agreed that they ought to give increased attention to the quality of lumber produced, since buyers became selective as times became harder. In addition, he thought it likely that the Ingram & Kennedy price schedules would require revision downward by June or July. But Ingram maintained that these concerns and worries were passing things. Demand would soon return to the lumber markets of the west, and in the meantime they were in a sound condition, fully able to withstand the rigors of financial adjustment. In the wake of wartime inflation, they had become securely established. Under the conditions then existing, they would be able to tolerate considerable reductions in price before margins disappeared.

Time would prove the correctness of the confidence expressed by Ingram in the spring of 1865, but the economic transition from war to peace caused hardship for some and uneasiness for many. Wheat prices had reached ninety cents in Wabasha toward the end of February, 1865, and seventy cents by the end of March. Randall observed that all of his farmer customers at Minneiska seemed to feel "Cross," and some of them "Sweare they will not farm another day & say they will not Pay the Present Prices for lumber. . . ."[38] Although the Minneiska manager was given to exaggera-

[37] Branch, Crooks & Company to Ingram & Kennedy, February 9, 1865.

[38] Gill to Ingram & Kennedy, February 25, 1865; Randall to Ingram & Kennedy, March 21, 1865.

tion, it was becoming increasingly apparent with the passing weeks that some early accommodation was necessary; either the price of wheat would have to go up or the price of lumber would have to come down. Since the first possibility was beyond the ability of lumbermen to control, there really was no choice in the matter.

In April, Gill reported that he had lowered the price on common lumber at the Wabasha yard by about $3, and that he had written Randall advising him to sell that grade for $26 per thousand at Minneiska. Randall's schedule of prices in late April was as follows: Common lumber and fencing, $26; clear lumber, $40; second clear lumber and flooring, $35; and shingles and lath, $6.[39] The prices were generally about a dollar less at the yard in Wabasha, and though down considerably from the peak levels of 1864, prices on lumber were still impressive.

List prices and actual sale prices, however, were often two very different figures, and before the first of May Gill acknowledged that he had sold a bill of common lumber for $20 per thousand. If this news was received at the Eau Claire office with some misgivings, it was received by Randall at Minneiska with downright resentment. Jabe had long imagined that he was operating under a definite handicap as concerned competition with Gill at Wabasha, both in lumber received and in prices he was permitted to offer. As his customers became fewer and the competition correspondingly greater, these suspicious assumed an added importance, particularly when Randall learned that prices at Wabasha were considerably below his own. Although Jabe Randall was not quite the astute businessman he claimed to be, in this instance his suspicions were not entirely without basis. Part of the problem was inherent in the system under which they operated, a system which allowed Gill greater leeway than Randall in the area of adjusting prices. This latitude could be critical when attempting to close a sale. But Randall was bound by instructions to sell at the price indicated on the schedule. Minneiska's schedule of prices came from the Eau Claire office by way of Wabasha, where Gill might up-date the figures as he thought necessary; and by the time the list reached Randall, they were never exactly current. This was not always a disadvantage.

[39] Gill to Ingram & Kennedy, April 22, 1865; Randall to Ingram & Kennedy, April 23, 1865.

During periods of rapid inflation, price lists from a superior or some central office seldom have an adverse effect on sales. Rather it is more likely that such schedules will increase sales, since prices are on the rise and the price stated on the schedule must be somewhat out-of-date, a bit behind, and lower than what could be charged at the time of sale. But the reverse becomes true when prices are on a steep decline. The salesman's schedule, still something less than current, tends to be higher in the prices it lists than the market will bear at that moment.

The solution to problems resulting from such a built-in time lag is to allow the salesman to make the required adjustments within a certain range. As noted, Gill was permitted this flexibility and he would often make changes as requested by Randall or as conditions demanded; but because the two yards were situated in such proximity, he might in the meantime have made a sale to one of the Minneiska customers. Not surprisingly, Randall's resentments increased as his sales decreased. He tried to make Ingram understand just how hard it was for him to "let a man go off with money in his Pants," repeating his contention, "I kinder think Gill wants Trade to Go to Wabasha for he does sell cheaper than he wants me to. . . ."[40]

Randall's complaints were just another indication to Ingram that business conditions were worsening. The end of good times unmasked all manner of difficulties, some real and some imagined. Hard times brought a return to hard competition, and all of the dissatisfactions that went with it. But as Ingram had already learned, the good years were to be enjoyed and the bad years were to be endured, and 1864 was no more typical a year than 1858 had been.

Ingram's perspective, however, did not eliminate the need to attend to some problems. The questions raised by the operation of the Ingram & Kennedy retail yards in southeastern Minnesota were typical of the kind of problem which Ingram was compelled to consider as his organization grew. This question was of fundamental importance, involving some provision for the flexibility necessary to encourage sound, timely business decisions at a number of different places, far removed from the home office and the direct control of the owners. It was obvious, if they were

[40] Randall to Ingram & Kennedy, May 9, 1865.

to continue their development, that they would have to depend increasingly upon trusted lieutenants to make the correct determinations. In the process, the orientation of the owners was necessarily shifted away from a primary concern with technological matters to an emphasis on managerial skills. The management of men was becoming as important as the making of lumber.

In truth, this adjustment was never made by some operators, and came only with difficulty to many. There were lumbermen who enjoyed, purely and simply, the mechanics of the lumbering operation. Desk work, even of the most pleasant variety, was to them akin to retirement. Accordingly, there were some firms whose size was strictly limited by the energy and talent of the owners. The corporate form of business organizations was alien to these individuals. They felt that when the business had reached the point where they could no longer exert direct personal control over all phases of its operation, then the business was too large. This was exactly the position that Kennedy would assume. It was the position that Ingram rejected, but none too easily.

In 1865, however, there were more immediate concerns than the probable need for organizational changes, concerns which centered on falling prices in the lumber market. Ethan Allen piloted the Mississippi rafts for Ingram & Kennedy during the season, and DeWitt C. Clark, later to be associated with Ingram in the banking business, served as agent in the downriver markets. In their efforts to dispose of the rafts to the best possible advantage, Allen and Clark encountered old and familiar problems. On one trip, for example, Clark declined a trade with Musser because he thought he would be able to do better elsewhere. Musser had offered $17.50 for the Ingram & Kennedy lumber in the raft and $17 for the Daniel Shaw Lumber Company cribs. In this instance Clark guessed wrong, and he ended up making a series of small, unprofitable sales, closing out what remained of the raft at Alton, Illinois, where he received $18 for the Ingram & Kennedy lumber and $17 for the Shaw Company cribs. Needless to say, the fifty-cent difference in price was insufficient to offset the additional costs incurred in running the raft from Muscatine to Alton.

Musser shared Clark's disappointment over their failure to agree upon a price, but in Muscatine he was experiencing some problems of his own, largely as a result of increasing amounts of

Chicago lumber in Iowa markets. Musser and other dealers fully appreciated that rate discrimination by the railroads in favor of Chicago operators was the primary source of their difficulties. A solution to the problem was less apparent. As he explained the situation, the railroads had to carry stock and grain to Chicago, and to return the cars loaded with lumber cost but little more than to bring them back empty. In any event, the effects of long-and-short-haul discrimination clearly complicated matters; and as Musser unhappily watched cars loaded with Chicago lumber passing through Muscatine for destinations farther west, he complained to Ingram that "this should not be."[41]

Musser's assessment did not halt the decline in lumber prices. By July of 1865, Ingram had ordered his agents to consider seriously any offer of $16 per thousand before continuing downriver. Meanwhile, Gill reported from Wabasha that although he was making every effort to keep retail prices on common lumber near the $20 level, he thought it probable that they would soon be selling that article at $16 per thousand feet.[42]

Lucius S. Fisher, who kept the books for Ingram & Kennedy at Eau Claire for a brief period in the mid-1860's, took a turn at downriver sales during the summer of 1865 with no exceptional success. Receiving reports in July that lumber was selling from $15 to $17 in the water at St. Louis, Ingram advised Fisher to close out the raft at Muscatine rather "than tie up or go farther." By that time, however, Musser was offering no more than $14 for the Ingram & Kennedy lumber in Fisher's raft, $13 for that of Daniel Shaw and $12 for the balance. Following the receipt of Ingram's memorandum, the only concession Fisher succeeded in getting from Musser was fifty cents more per thousand for the Ingram & Kennedy cribs, and the trade was closed at those low figures.[43]

One of the major advantages enjoyed by Ingram & Kennedy, especially during hard times, was the quality of their lumber.

---

[41] R. Musser to Ingram, May 27, 1865. Musser cited the specific example of freight rates to Grinnell, Iowa, located some one hundred miles west of Muscatine and three times that distance from Chicago. The lumber rate per thousand feet, Muscatine to Grinnell, was $9.00; from Chicago to Grinnell it was only $13.45. Musser to Ingram & Kennedy, May 25, 1865.

[42] Gill to Ingram & Kennedy, July 8, 1865.

[43] Fisher to Ingram & Kennedy, July 17, 1865.

Indeed, there was less concern with careful sorting when demand was large, buyers then tending to be less particular and less given to complaint. In the near future, when quantity of production would become the crucial consideration, when all lumber would be cheap and the cost per unit more critical than quality, Ingram & Kennedy would make the corresponding adjustments. This was not true of all firms. For example, the Daniel Shaw Company apparently maintained an effort to produce high-quality lumber even after business conditions actually discouraged such concern, when prices received did not justify the effort. But at least until the late 1860's, with a reputation yet to establish, the business of Ingram & Kennedy was not so large that the knowledge and skill of the partners failed to show in the product of their sawmill.

When buyers became increasingly selective, as they did during the summer of 1865, it was quality that determined sales. Thus when Peterson & Robb, dealers at Dubuque, informed DeWitt C. Clark that the Ingram & Kennedy lumber was the "best pulled here this season," it was more than just complimentary. The important result was that Peterson & Robb immediately ordered another million feet.

One of the minor complications relating to this question of quality involved the continued necessity of rafting lumber from various sawmills together. As yet there were few operators in the Chippewa valley who produced lumber in sufficient quantity to enable them to fill out, with any convenience, an entire raft for Mississippi running. Therefore, two or more mills would combine their output into a single raft. Largely because of their proximity, Ingram & Kennedy and Daniel Shaw ran their lumber together for many seasons, but it was not uncommon for cribs from other operators, such as Stephen Marston and Stanley & Skinner of Chippewa Falls, to make up a portion of the raft. Naturally, the job of selling was made more difficult because of the varying qualities produced by the different mills. At one point a Muscatine dealer advised Ingram not to "injure the Sale of your own make by having others of poorer quality with it."[44] This advice would be followed as soon as the level of production permitted.

[44] S. G. Stein to Ingram, January 3, 1866.

But it was hoped that worries such as these, in large measure the product of hard times, would shortly diminish in importance. That possibility seemed not unlikely as demand and prices for lumber stiffened and then began to improve in the late summer and early fall of 1865. The effects of the resumption of railroad construction and emigration into the trans-Mississippi West were again becoming evident. The change came rather suddenly. Musser offered $19 per thousand for a raft to be delivered him in September. Ethan Allen, the raft pilot, recognized the change in conditions from the increased number of signals he received to stop at various river towns. Even Jabe Randall noted an improvement at Minneiska. He had been sick for quite a while, but the activity in the little yard revived him considerably, and he looked forward to "a Big Trade this next Month Provided I can have what [lumber] I want."[45] The arrival of wintry weather did not discourage customers to any significant extent, Gill selling 140,000 feet at Wabasha during the month of November with Randall doing comparably as well at Minneiska.

The immediate outlook, however, was not without its darker side. Despite the revival in the lumber market, wheat remained a low item. With winter beginning to settle in, farmers were not inclined to sell at eighty and eighty-five cents a bushel. Randall reported that his farmer customers all wanted lumber and many had begun to inquire whether the yard might be willing to accept wheat receipts for security on lumber. Jabe, who liked to do business as much as anyone, was not tempted in the least to make sales in such a manner, not yet anyway. As he confided to Ingram on Christmas Day: "Their [sic] is not Ten Farmers in 50 that will do as they agree, Take them as a General Rule."[46] Nevertheless, without such accommodation and lacking any other arrangements, if grain did not move, lumber moved slowly.

But during the winter months sales were expected to lag, and despite the usual assortment of fears for the future, the situation of Ingram & Kennedy at the end of 1865 seemed at least moderately encouraging. The war years had literally put them on their feet, and though there would doubtless be hard times ahead, it seemed unlikely that such times could be as difficult as

[45] Randall to Ingram & Kennedy, September 28, 1865.
[46] Randall to Ingram, December 25, 1865.

those already endured. If much remained to be done, it was entirely understandable that Ingram found the winter of 1865–1866 an opportune time for a trip back home. He was anxious to assure friends and relatives around Lake George and along the Ottawa that the Ingrams were happy and in good health and, perhaps of greatest interest and importance, that Orrin Ingram was a success. The impertinent Jabe Randall may have been close to the mark when he addressed Kennedy during Ingram's absence, "I presume O. H. is having a Good Time," adding, with obvious admiration, "he is just the man that knows how to Enjoy himself & no wonder, with the name, Fame & the Clink you are able to carry."[47]

Randall tended to be overly impressed by such considerations, but it was increasingly evident to the citizens of Eau Claire that their neighbors, the Ingrams and the Kennedys, were fast becoming affluent. If they had not observed the progress previously, they could hardly have overlooked a list of reported incomes for 1866 which Henry C. Putnam, then an assistant assessor with the federal office of internal revenue, thoughtfully provided the local paper. There were a large number of Eau Claire's elite admitting to incomes in excess of $1000. Ingram and Kennedy each reported $7,750, ranking them third on the list—considerably below Joseph G. Thorp's $32,075, but impressive nonetheless.[48]

Having endured the immediate postwar deflationary adjustments with minor inconvenience, the partners settled down to the business of making as much lumber as they could as cheaply as they could. The rapidly growing demand in the markets of the west assured the sale of all lumber produced; whether or not profits resulted depended upon the costs of production and transportation. If the lumbermen of this period had a complaint, it was the low prices of lumber. For the next ten years or so, rates remained surprisingly stable, seldom exceeding $15 per thousand feet or falling below $12 for common boards in the Mississippi River markets at Hannibal and above. But such prices were not so low as to deny a reasonable margin to the efficient producer.

Although it was a lesson easily forgotten, the kind of lumber most in demand was cheap lumber. Quantity, not quality, was

[47] Randall to Ingram & Kennedy, January 17, 1866.
[48] *Eau Claire Free Press,* June 15, 1867.

rapidly becoming the key to successful production efforts. Whether boards carried the marks of the saw, were sap-stained or splintered, were of irregular lengths and uneven edges was of less importance than the fact that they were mass-produced at small expense. While hardly a luxury, lumber could still price itself out of a market made up largely of pioneer farmers, who were chronically hard-pressed for ready cash. Faced with these circumstances, lumbermen built new mills and enlarged old mills, concentrating their efforts on increasing sawing capacities and speed and efficiency in the handling of logs and lumber.

As correct as the orientation of these activities might have been, lumbermen could not avoid giving some attention to other phases of their operation. Ingram, for example, was beginning to appreciate many of the effects of his company's recent and rapid growth. One of the more apparent changes occurring in the course of the inflation brought on by war was that he was now in a position to provide credit as well as consume credit. Many other producers found themselves in a similar situation, thereby increasing the flexibility of operation and, concurrently, the opportunities for greater competition. This would shortly prove important within the valley of the Chippewa, when outside producers sought access to the pine resources of that region.

From this new ability to provide credit, even in very limited amounts, stemmed a basic question relating to the marketing phase which was becoming more pressing with the passing months. It involved the subject of cash or credit sales. The traditional tendency had been to insist on cash payment, particularly in retail dealings, with an accompanying assumption that those who requested time were less than honest. Gradually, however, in both retail and wholesale activities, Ingram came to accept the fact that to require "cash on the barrelhead" was not always sound policy. But the acceptance of this change did not come easily, and certainly not at the Minneiska yard of Ingram & Kennedy. There Jabe Randall, after only a brief experience with allowing time on lumber sold, concluded that he was being duped by his farmer customers. "Darn them," Randall complained, as a general practice they "want to Run a year in advance on Tick for Evrything & when they Git any money then they want to buy all the Land Joining them."[49]

---

[49] Randall to Ingram & Kennedy, September 10, 1867.

Indeed, there were some real disadvantages inherent in selling on credit. Even when operating on a strictly cash basis, the lumberman had considerable capital tied up for extended periods in land, logs, and lumber, as well as in the sawmill and other facilities. Accordingly, any change which would increase the accounts receivable was not readily accepted. Whatever the disadvantages of credit sales, however, they were slight when compared to the disadvantages of no sales. Also, there were some benefits to be gained from selling on time. For example, through various devices, prices on credit sales were normally higher than was the case with cash trades. Gill admitted in March, 1867, almost apologetically, that he had been selling some lumber on time at Wabasha "till after Harvest" for $25 a thousand feet, plus 10 per cent interest. Even without the interest, $25 was $2 more per thousand than his cash retail price.[50] In later years, as credit sales became an accepted practice, the more common adjustment was a discount extended to those paying cash.

Although the question of cash or credit sales was of basic importance, it was one for which there was no final answer, always depending in the end on someone's judgment in an ever-varying set of circumstances. A far more definite question concerning marketing arrangements was posed by Major William H. Day of Dubuque, Iowa, in the spring of 1867. Major Day suggested that Ingram & Kennedy open a wholesale yard at Dubuque, allowing him to take an interest and to manage the operation. Day's experience in lumber wholesaling was quite limited, but Ingram was much impressed with his enthusiasm and his business sense, and he persuaded Donald Kennedy that they would do well to accept this proposition from Dubuque.

Ingram & Kennedy no longer had the same critical need for business connections with eastern friends and financiers, except in the case of large expenditures such as the purchase of pine land. Consequently, western opportunities for association were becoming increasingly attractive. Still, acceptance of the offer advanced by Major Day was a difficult decision for the young firm to make. In a sense the partners had reached the fork in the road. They now had to determine whether they were in business to make a living or to make a fortune. Kennedy clearly had

[50] Gill to Ingram & Kennedy, March 16, 1867.

his doubts, but such was not the case with Ingram. Although a separate wholesale operation and involvement with a new partner might commit the firm to further expansion of production and related activities, Ingram did not hesitate. The arrangements were completed in time for the partnership of Ingram, Kennedy & Day to begin selling lumber with the start of rafting in early May, 1867.[51] Prices that spring opened at $19 per thousand in Dubuque, and Day looked forward to busy times, noting that the railroad was soon to be completed as far west as Ft. Dodge, Iowa, "which in itself will create a largely increased demand." Because of the growth of the market area, and for many other reasons, Ingram, Kennedy & Day proved a lasting and most profitable organization.

Initially, the Major's primary expressed concern was that his new partners would fail to recognize the importance of supplying the Dubuque yard strictly with lumber cut from "Chipeway logs." This distinction reflected a common preference for pine from the mainstream or upper basin of the Chippewa as opposed to logs cut from pineries along the Eau Claire River tributary. Day advised that the Dubuque market required a higher percentage of upper grades, especially during the winter and spring, than could be provided by lumber sawed out of the rougher Eau Claire River logs.[52] This difference between sawlogs was so widely and firmly recognized that it must have been more than mere prejudice. For example, later that same season at Wabasha, a market never known as hard to please, Gill requested a raft of lumber from which to make flooring. He specified Chippewa logs, because, he asserted, Eau Claire River lumber would "not give . . . anough [sic]."[53]

There were real differences between pineries in the quality of logs cut, with a corresponding variance in the quality of lumber produced, but the much-discussed market differences were of little actual importance. For all the talk, Ingram & Kennedy made few adjustments to satisfy buyers in any of the Mississippi River markets. On those occasions when the percentage

<hr />

[51] Major Day put up one-third, or $3000, of the initial investment. Playter to Ingram & Kennedy, May 16, 1867. See also Day to Ingram, March 18, 1867.

[52] Day to Ingram & Kennedy, May 27, 1867.

[53] Gill to Ingram & Kennedy, October 12, 1867.

of this kind or that grade caused complaint, a slight adjustment in price always seemed to provide relief. In any event, the firm sold considerably more fencing than flooring, and many more ordinary barn boards than No. 1 clear pine.

The first season for the Dubuque yard turned out very well. Although prices on common lumber fell to $15 by September, the volume of business handled by Ingram, Kennedy & Day remained large. Throughout the fall of 1867, Day was able to forward a thousand dollars to the Eau Claire office nearly every other week. Appreciated almost as much as the cash, however, were the increased opportunities for credit afforded by the Dubuque partnership through Day's account with the Merchants National Bank. He succeeded in making arrangements by which they could borrow a few thousand dollars without any prior notice, though Day tended to be much more reluctant than Ingram "to use our bank credit when not necessary."[54]

Lumber prices throughout the west had fallen to unsatisfactory levels, at least according to producers. Agent Playter had managed to sell some Ingram & Kennedy lumber for $15.50 from the water at Davenport in early October. Generally, however, the prices late in the 1867 season were weak at $15, causing some anxiety among operators who selectively recalled the days of ten-dollar sales. At the Dubuque yard, Major Day hesitated to predict whether the million feet then in stock would prove to be enough or too much for the winter and early spring trade. While admitting that business of late had been "weak & drooping," Day maintained that the markets were not over-supplied, and for all of the complaining about low prices, he noted that there was "still a fair margin of profit between yard & water prices."[55]

Indeed, Major Day gave early evidence of an attitude and an approach that was to prove a real advantage in the lumber business. Perhaps even more than his partners, Day was convinced that profitable operation depended almost entirely on volume of trade; that it was not to be measured by the price per thousand board feet but by the number of thousand feet sold. His plans and policies reflected these attitudes. Consequently, the Dubuque yard was not simply going to be a place to stockpile lumber

[54] Day to Ingram, September 28 and October 2, 1867.
[55] Day to Ingram & Kennedy, October 31 and November 5, 1867.

which could not be sold for profit in the water, at least not as long as Major Day was the manager. He was ambitious and he was bright, and if he and Ingram occasionally disagreed over policy questions, Ingram refused to allow any such differences to obscure the mutual benefits to be gained by continued association. Although there was at least one instance when Day threatened to quit the company as a result of differences in opinion with Ingram, compromise carried them through that and other periods of minor turbulence. Ingram and the Major were each too strong and full of ideas not to have their disagreements, but they remained the closest of partners so long as there was pine standing in the valley of the Chippewa and beyond in other valleys.

In a real sense, Day forced expansion on Ingram at a rate faster than otherwise would have occurred. By the summer of 1868, Day was insisting that they needed more room for the piling of lumber at Dubuque and additional property was purchased, on time, to the extent of $7000. Ingram, Kennedy & Day also opened a little retail yard that summer at Winthrop, Iowa, sixty miles west of Dubuque, the first of many such operations. While these retail yards were not always successful in that they returned a profit to Eau Claire via Dubuque, they did provide some of the same flexibility to the Dubuque facility of Ingram, Kennedy & Day in the handling of its lumber that the yards at Wabasha, Minneiska, and Dubuque provided the Eau Claire sawmill of Ingram & Kennedy.

Policy followed by the Major required that the Iowa yard managers be in partnership with Ingram, Kennedy & Day, and whether a retail yard was successful depended in large measure on who was in charge. Insistence on an investment by the managers may have had something to do with security, or even with a recognition of certain profit-sharing advantages; but more commonly, a man's ability to invest was considered a badge of respectability Indeed, the first step in the process of establishing a retail yard was finding an individual interested in selling lumber who had some idle funds. Selection of a likely locality was less a problem. Having money to invest in such a yard, however, was no guarantee of business ability or honesty, as Day and Ingram discovered more than once.

Lumber prices in the Dubuque market had remained steady

throughout most of the 1868 season. Nevertheless, Ingram, Kennedy & Day continued to do considerable business at $15 dollars per thousand feet. By the middle of June, Major Day was looking forward to the fall and winter trade, advising that they should be piling at an increased rate, sufficient to assure the receipt of at least a million feet in each of the next two months. He was still demanding more lumber in September when Ingram intervened with firm disapproval. The problem with the yarding of lumber was that payment was postponed. Day shortly was made to understand that a great many demands were then being made on the resources of Ingram & Kennedy. Thus, while he continued to iterate his desire to yard as much as possible, he also acknowledged that his partners' investment in Dubuque was already quite large, "and your wants above are probably proportionately large, & it is quite probable that we cannot meet them without some further water sales."[56]

In any event, the Major had been overly optimistic concerning the demand for lumber. The fall of 1868 proved a slack period in the Mississippi River lumber markets, and the winter which followed was similarly unrushed. Part of the reason for the slow sales had to do with wheat prices, which were again in a period of decline. In early November, Major Day reported that farmers were receiving seventy-five cents a bushel for their wheat in Dubuque, a reduction of more than one-half from price levels immediately following harvest. As was their accustomed practice, farmers held their grain "for higher prices in almost every instance where by any means they are able to do so."[57] Similar conditions prevailed in southeastern Minnesota, where Gill reported that his farmer customers were again requesting lumber on wheat —their way of asking for credit. Assuming that the inactivity at the Wabasha yard was only a temporary thing, Gill initially replied to the credit-seekers that he was through giving them time on lumber: 'It is Cash up or no trade here now."[58] When trade continued dull, however, Gill became rather more accommodating.

Aside from the consequences of small demand and any accompanying miscalculations as to the amount of lumber required, it was not possible for Ingram and his yard managers always to

---

[56] Day to Ingram, August 22 and September 11, 1868.
[57] Day to Ingram, November 5, 1868.
[58] Gill to Ingram & Kennedy, November 7, 1868.

march in step. For example, Day's objectives at Dubuque may
have been identical with those of Ingram, but the Dubuque wholesale operation remained the focus of the Major's orientation while
Ingram had to be concerned with matters of supply, processing,
and other outlets in addition to Ingram, Kennedy & Day. Ingram was also trying to reorganize above Dubuque, a problem
Day neither understood nor cared much about.

Concurrent with the establishment of Ingram, Kennedy & Day
in Dubuque, Ingram had undertaken the task of bringing greater efficiency to the operations at Wabasha and Minneiska. Primarily this involved some changes in management of those yards.
One of these concerned Jabe Randall who had experienced a
rather poor year in 1867 although, as he maintained, it was not
through any fault of his own. "I am shure I try hard enough,"
he told his employers, with the added assurance that "if the
thing was my own, I would not take more Intrest." But no one
seemed too surprised or disappointed when Jabe decided during
the following winter that he had had enough of the retail lumber
business and wanted to be replaced, explaining, "my health is
not vary Good & [I] want to Go up to Eau Claire . . . & Rusticate
Some."[59] With abnormal haste, Randall was relieved by Albert
J. Archer, another Canadian; and if the correspondence with
Minneiska was no longer quite so quaint, the business of the yard
did not suffer as a result of the change.

No such personnel problems existed at Wabasha. In fact,
William Gill had proven such a trusted and valued employee
that the partnership of Ingram Kennedy & Gill, established in
1867, seemed the logical consequence. Similar to the organization at Dubuque, though on a much smaller scale, Ingram, Kennedy & Gill had plans for expansion and looked forward to business at a brisker rate and larger volume. In addition to the yards
at Wabasha and Minneiska, a smaller retail outlet was opened at
nearby Kellogg. By the fall of 1867, work had been completed on
a planing mill at Wabasha and shortly afterward a steam drying
house or dry kiln had been added to the facilities, enabling them
to sell smooth and seasoned finishing boards at significantly higher rates than the rough stock. By 1870, a sash-and-door factory was
operating in connection with the yard at Wabasha, supplying win-

[59] Randall to Ingram & Kennedy, January 14, 1868.

dows and doors to the three local yards and to all other Ingram & Kennedy outlets as well. But this was about as far as Ingram, Kennedy & Gill could reasonably grow since their marketing area in southeastern Minnesota, although unusually secure, was strictly limited. Further expansion would encounter stiff competition from Stillwater, Minneapolis, or Winona producers, depending on the direction of that expansion. So the organization remained substantially content to maintain their business in such a manner so as to permit continued monopoly over their small but stable market.

Ingram, Kennedy & Day, however, was not similarly restricted, and the plans for that operation, especially as proposed by Major Day at Dubuque, envisioned early and large-scale extension of activities. During the quiet days of the winter of 1868–1869, bothered by too few customers, Day had plenty of opportunity to contemplate just how the position of Ingram, Kennedy & Day might best be improved. The essential result of these deliberations involved the construction of a sawmill at Dubuque.

There were a number of good reasons supporting such a project. In the first place, Ingram & Kennedy logs often escaped the booms at Eau Claire, requiring that downriver arrangements had to be made, usually at little or no advantage. It was true also that experience had proved the difficulty of keeping the yard inventory of Ingram, Kennedy & Day complete in all grades and kinds, with the result that to fill certain bills Major Day had no choice other than using ready cash to purchase those articles from neighboring competitors. Finally, at least during good times, it was apparent that demand for Ingram & Kennedy lumber far exceeded the capacity of the Eau Claire sawmilling facilities to satisfy the growing requirements of their own yards as well as the requests of their wholesale customers.

By the end of January, Day's deliberations had progressed to the point that he was considering and comparing the log-holding capacities of various sloughs with a view towards selecting the exact location for the new mill. When he reported the details of his survey to the Eau Claire office, however, the Major found enthusiasm lacking. Without bothering to discuss the possible merits of the plan, Ingram simply informed Day that his efforts were somewhat premature. As Ingram viewed their situation, they were not yet in condition to consider an increase in the

Dubuque investment, at least not to the extent of constructing a sawmill.[60]

From the vantage of Dubuque, the proposal offered by Major Day appeared entirely reasonable. But Ingram had begun to appreciate that sawlog supply would likely be a problem, and shortly too; he was becoming increasingly concerned with the task of keeping the Eau Claire mills adequately stocked with logs. The thought of providing for another mill at Dubuque was less than agreeable. The problem was immediately complicated by Ingram's current negotiations for the purchase of the Sherman Lumber Company's Eddy Mill, located some distance upstream from Half Moon Lake. As a further and most important consideration, involving the question of future pine resources, there was the nagging uncertainty created by the entry into the Chippewa valley of the so-called Beef Slough interests. Would they be successful, and if so, to what extent? Ingram had been deliberately cautious in the purchase of pine land, but he was becoming increasingly less certain that this policy could long be continued. Although he hoped that future success in lumbering would not be dependent on the ownership of large tracts of timber, he was not sure.

Ingram had received more than just a hint of future developments in early February, 1867. State Senator Joseph G. Thorp had written him from Madison upon learning there was soon to be an application made for the construction of a boom at Beef Slough, observing, "that won't do—too much water runs [in] there now and we don't want to run logs in there and deepen it—rather close it up." But their concern was more than hydrographic, and Ingram had no need for Thorp's prompting that "Other important reasons will present themselves to your mind against it. . . ."[61]

Ingram had worried for some time that the growing demand for lumber would so increase the value of standing pine that fierce and costly competition would be introduced into the forests of the Chippewa valley, and that the procurement of sawlogs rather than the production of lumber would become the predomi-

[60] Day to Ingram, January 17, 1869.
[61] Thorp to Ingram, February 6, 1867.

nant element in the lumbering process. Perhaps basic to these fears was the realization that the possession of large amounts of capital would be of greater importance in the future than knowledge, skill, and experience in lumbering. Anyone with money could purchase pine land; and although he had enjoyed relative success, Ingram could not foresee the day when he and Kennedy would be able to compete in a contest so entirely financial.

The contents of Thorp's letter provided the clearest evidence yet that what had long been feared was nearly fact. In a real sense it introduced the most crucial period in the lumber industry of the Chippewa valley, and certainly the most colorful. For more than the decade which followed, a struggle for control of the river and the vast pine resources within the basin would dominate the interests and engage the energies of lumbermen at Eau Claire, Chippewa Falls, and many points downstream on the Mississippi.

Although not strictly two-sided, the contest is most easily understood as one between local and downriver millmen. The Eau Claire, Chippewa Falls, and other resident Wisconsin operators viewed the pine of the Chippewa valley as their own and considered the opportunity for processing it as strictly a local prerogative. In opposition were the nonresidents. Some were the actual owners of tracts of pine land in the valley, interested in selling timber and logs to whoever would pay the highest prices. Others were the owners of sawmills downriver on the Mississippi who, lacking forests of their own, viewed the Chippewa pineries as simply one of many sources of logs. But the valley of the Chippewa was considerably more than just one of many sources—it was to be the largest single source of logs for the sawmills of Iowa, Illinois, and Missouri.

This pre-eminence was due to a number of factors. In the first place, the Chippewa River and its tributaries drained a very large area, nearly one-sixth of Wisconsin and more than one-third of the region containing pine. Merely from a quantitative consideration, the Chippewa valley was special. Also, comparatively little of the pine had been cut by the end of the Civil War, and indeed not much was even owned by lumbermen at that time. With such opportunities for investment, Chippewa pine lands attracted increasing attention with the result that by 1867, individuals such as Detroit financier Francis Palms and institutions such as Cornell University had expended money and scrip to acquire large and valuable tracts of timber.

Naturally, the resident lumbermen viewed these entries with misgivings, but they lacked the funds to make such purchases themselves and by that means deny the opportunity to outsiders. One of the early advantages of the Chippewa millmen was the abundance of pine, which permitted firms like Ingram & Kennedy to ignore or at least to postpone timberland purchase. Accordingly, they could expend their very limited capital on more immediate requirements. Finally, as noted earlier, the streams and rivers of the Chippewa valley provided an unequaled transportation network whose waters flowed in the desired direction, enabling the logs to be obtained at relatively small expense.

Because of this combination of factors—the amount of standing pine, the availability of the pine lands, and accessibility of the pine—during the post-Civil War period this region became the focal point for sawmill operators in search of a dependable source of logs. Defending these forests as best they could were the resident lumbermen, who understood that more than half the value of the pine would be lost if the logs were sawed into lumber elsewhere. Although they were unable to prevent the entry of nonresidents into their forests, for a time they felt that they might nevertheless prevent the exit of pine logs on their streams.

It was this hope which brought sudden notoriety in the late 1860's to a slow-moving, meandering stretch of water known as Beef Slough. Like many rivers, the Chippewa fans out near its mouth, its waters seeking passage through the alluvium and following a variety of routes to the Mississippi mainstream. Beef Slough was one such channel of the Chippewa. If millmen along the Mississippi River were ever to succeed in getting logs out of the Chippewa pineries, their first requirement was a secure place in which to catch, sort, and scale the logs and construct the rafts. Beef Slough was ideally suited and situated for that purpose, and its crucial importance was lost neither on the "outsiders" interested in running logs out of the Chippewa valley nor on the local lumbermen dedicated to the prevention of such an exodus.

The contest which grew out of these conflicting objectives has been much described and discussed. Briefly, the local interests initially sought to deny the nonresidents use of Beef Slough by obtaining a franchise from the state legislature which reserved the booming privileges at the Slough for themselves.[62] A crack in the defenses early appeared as James H. Bacon, a recent arrival from Michigan who represented the "outside" interests of financier Francis Palms, somehow managed to get his name included in the list of local incorporators.[63] The error was soon discovered, however, and within two weeks the 1866 legislature had amended the original act, with the name of Bacon a notable deletion.[64] No doubt as a result of this unexplained bit of confusion, by means of an 1866 amendment the Chippewa valley millmen dusted off an 1856 franchise which granted to the incorporators of the Chippewa River Improvement Company "full power to dam all bayous and sloughs that make out of the Chippewa river, and to confine said river to its proper channel."[65] The obvious advantage of this old charter resulted from the fact that while the 1856 act mentioned almost every pioneer lumberman in the Chippewa valley, there was then no James H. Bacon to muddy the waters. The resident operators clearly did not intend to *use* the privileges granted by the franchises, old or new. At the same time, however, they sought to *reserve* those privileges for themselves, thereby denying them to others interested in floating logs out of the Chippewa.

But these and a great many similar maneuverings proved, in the end, to be insufficient. The "outsiders" succeeded in incorporating their Beef Slough Manufacturing, Booming, Log-Driving and Transportation Company under the laws of Wisconsin. Indeed, this step had not been seriously contested since it was assumed that a company lacking the authority necessary for its operation could pose no threat.[66] In fact the nonresidents were not successful in their efforts to obtain a franchise for privileges at Beef Slough, but this series of legislative defeats had little or no effect on their determination to proceed with the plans. Com-

[62] *Journal of the Wisconsin State Senate,* 1866, p. 314.

[63] *Private & Local Laws of Wisconsin,* 1866, Chapter 181.

[64] *Private & Local Laws of Wisconsin,* 1866, Chapter 286.

[65] *Private & Local Laws of Wisconsin,* 1866, Chapter 509; 1856, Chapter 405.

[66] *Revised Statutes of Wisconsin,* 1867, Chapter 73; Thomas E. Randall, *History of Chippewa Valley* (Eau Claire, Wisconsin: Free Press Print, 1875), 142; George W. Hotchkiss, *History of the Lumber and Forest Industry of the Northwest* (Chicago: George W. Hotchkiss & Company, 1898), 181.

monly known as the Beef Slough Company, headquarters were established in 1867 at Alma, just below the mouth of the strategic channel, and superintendent Timothy Crane was directed to construct facilities for use during the following season.

The local operators were fast realizing that it would require more than defeats in the legislature to discourage the Beef Slough interests. Thus, acting under the authority of the Chippewa River Improvement Company franchise, Eau Claire lumbermen and representatives of Knapp, Stout & Company of Menomonie decided it was now time to "do some work" at the slough themselves. In what amounted to the simplest of improvements, the entrance to Beef Slough was closed by a rough but effective dam—under the pretext that the dam would assist in confining the Chippewa to its proper channel.

Although opinions varied in the extreme as to whether so direct an action was the work of law-abiding businessmen or of a vigilante committee, it was apparent to all that the urgency of the undertaking at Beef Slough had no natural origin. But this did not prevent the *Eau Claire Free Press* from placing its account of the affair under the headline, GRAND IMPROVEMENT OF NAVIGATION, the editor cheerfully reporting that the dam had resulted in an immediate difference "of several inches of water on this extremely shallow and bad portion of the river."[67] The editor of the *Chippewa Union and Times* of Chippewa Falls, however, adopted a somewhat different view. He described the recent activities of "the Eau Claire and Menomonie saw mill men as the most dastardly, high handed, outrageous piece of business ever undertaken by men."[68]

The *Alma Journal,* by then the journalistic spokesman of the Beef Slough interests and known in Eau Claire as the *"Alma Journal and Beef Slough Advocate,"* predictably assailed the dam construction in no uncertain terms. Furthermore, its editor maintained, Beef Slough was, in fact, not a slough but a navigable stream which, if not efficient in its progression toward confluence with the Mississippi, had as much right to meander as any other navigable stream.[69]

[67] *Eau Claire Free Press,* August 1, 1867.
[68] *Chippewa Union and Times,* June 26, 1867.
[69] *Alma Journal,* August 7, 1867.

The local forces had left a detachment of guards behind on "the little Beef Dam" for purposes of protecting it against any possible contingency. But despite the guards' assurance that "one of us never Sturs [*sic*] off the dam,"[70] in late August some of the Beef Slough Company employees tore out the dam improvement, and the waters flowed much as before. Now the editor of the *Eau Claire Free Press* could take his turn at crying foul. This he did, offering the destruction of the Chippewa River Improvement Company dam as clear evidence that the Beef Slough interests were dedicated "to accomplish by mob violence, what they failed to achieve by law, and are engaged in carrying out their hellish and unlawful proceedings." He further warned that "this posse of law-breakers" would be back in Madison in the spring, seeking to legitimize their intrigues, predicting the direst of consequences should that "mob-monopoly" be successful at the next session of the legislature.[71] Indeed, the scene of battle did return to the somewhat more decorous halls of the Wisconsin capitol.

Orrin Ingram, along with other valley lumbermen, spent considerable time and money during the winter of 1867–1868 in an effort to cement their Chippewa political alliance, thereby assuring the continued frustration of the Beef Slough Company objectives in the state legislature. These efforts appeared to have been successful when the Beef Slough Charter bill went down to decisive defeat in the Wisconsin Assembly on February 28. But as a few had feared, final and complete victory required more than mere voting strength. Senator Thorp had warned that constant vigilance was also necessary, and on that count the Chippewa alliance proved deficient. In fact, residents of Eau Claire had almost no opportunity to savor the headlines in the *Free Press* announcing BEEF SLOUGH MONOPOLISTS VOTED DOWN IN THE ASSEMBLY, before they were confronted with a new crisis.[72]

The headlines were correct. The Beef Slough bill had been voted down, but when James Bacon arrived in Eau Claire on March 9, it was not long before Ingram and his neighbors realized that all was not right. Looking anything but a loser, "Beef Slough Bacon" smugly revealed that by means of a clever strata-

---

[70] Reinelt D. McDaniels to Ingram, August 2, 1867.

[71] *Eau Claire Free Press*, August 29, 1867.

[72] *Ibid.*, March 5, 1868.

gem his company had achieved their legislative objectives despite their apparent defeat.

The details soon became known. Closely following the Assembly rejection of the Beef Slough Charter measure, a bill "to incorporate the Portage City Gas Light Company" routinely passed both houses of the Wisconsin legislature with a rider which had been tacked on by Senator Moses M. Davis, who, not incidentally, was a member of the Beef Slough organization.[73] This rider—section nine of the bill, unnoticed by those who should have been alert—had been carefully drawn for purposes far removed from a power company for Portage. The primary consideration of the Beef Slough interests was the fact that James H. Bacon had been included in the list of incorporators of the original Beef Slough charter in 1866. The rider therefore provided that when a franchise had been granted by law to several persons, the grant would be deemed several as well as joint, "so that one or more may accept and exercise the franchise alone."[74]

In essence this meant that since the rights granted at Beef Slough in the 1866 charter had not been exercised, it was now legal for any of the individual incorporators to exercise those rights if he so desired. Although obviously in error, James H. Bacon had been listed as one of the original incorporators, and he now announced his intention to assign his rights over to the Beef Slough Manufacturing, Booming, Log-Driving and Transportation Company.[75] The deception successfully executed by Senator Davis, however, did not end the contest. Rather the focus

---

[73] *Journal of the Wisconsin State Assembly,* 1868, pp. 89, 949; *Journal of the Wisconsin State Senate,* 1868, p. 544.

[74] *Private & Local Laws,* 1868, Chapter 331.

[75] Merk, *Economic History of Wisconsin During the Civil War Decade,* 94; John C. Gregory, editor, *West Central Wisconsin: A History* (Indianapolis: S. J. Clarke Company, Inc., 1933), II: 828; Robert F. Fries, *Empire in Pine* (Madison: State Historical Society of Wisconsin, 1951), 143–144; James Willard Hurst, *Law and Economic Growth: The Legal History of the Lumber Industry in Wisconsin, 1836–1915* (Cambridge: Balknap Press of Harvard University Press, 1964), 267; Hidy, *et al., Timber and Men,* 47. Bacon is a shadowy figure throughout. Never anything but an outsider in the valley, it is an over-simplification to state, as Hurst and others have done, that he "switched allegiance." Rather the real puzzle remains as to how he managed to get his name included in the original list of incorporators. He was a native of Ypsilanti, Michigan, and came to Wisconsin to care for and watch over the business interests of Francis Palms. Although he was important for a time, Bacon left few tracks.

of attention and activity merely moved again from Madison back to the Chippewa.

As the opposing groups well understood, in order for any logs ever to reach the quiet waters of Beef Slough, they would first have to pass through the booms of every sawmill on the river. That the Chippewa valley operators would give any priority to the logs bearing Beef Slough Company marks was out of the question, although in 1868 these millmen could contend, with some degree of sincerity, that they lacked the facilities required to insure the rapid sorting and passage of such large numbers of logs. In any event, assuming his responsibilities as manager of the Beef Slough drive, James Bacon realized that he could not trust the success of his operation to good will and gravity. Accordingly, he advised Orrin Ingram and the other valley lumbermen that the Beef Slough Company expected all of its logs to be passed downstream without delay. Implicit in the announcement was the threat that company drivers would be the judges of what constituted delay.[76] Such a prospect was disquieting to the resident operators.

The Beef Slough War began when the ice went out and the logs commenced their downstream journey. Although violence was limited to a couple of picturesque "affrays," for those who had money invested in Chippewa sawlogs it was a period of great anxiety. Few operators could bear the losses which would result from their booms being forcibly opened, allowing all of their logs to escape below. To Ingram the crisis seemed sufficiently serious that for one of the rare times in his life he worked on the Sabbath. Armed with pike poles, he and Kennedy, together with Daniel Shaw, Buffington, Marston, and other owners, put in long hours shoving as many logs as they could through the canal leading into Half Moon Lake and out of the path of the Beef Slough drive.[77]

Despite all of the effort, excitement, and expense, by the end of May the *Alma Journal* reported that only 15 million feet of logs had reached the safe confines of the slough, and that estimate was doubtless an exaggeration.[78] Clearly the total received

[76] Randall, *Chippewa Valley*, 143.
[77] Bailey, ed., *Eau Claire County*, 366–367.
[78] Hidy, *et al., Timber and Men*, 48, states that only 12 million feet of logs reached Beef Slough in 1868.

was far below the 50 million feet for which the Beef Slough Company had contracted to deliver in this, their first season of operation. But of greater long-range significance, while it was obviously not a financial success, the Beef Slough drive of 1868 established the precedent. From that season forward, the pineries of the Chippewa valley would furnish an ever-increasing quantity of logs to the Mississippi River sawmills. Included in the quota, beginning in 1868, were logs contracted by Frederick Weyerhaeuser and Frederick C. A. Denkmann of Rock Island, Illinois.[79]

But the stockholders of the Beef Slough Manufacturing, Booming, Log-Driving and Transportation Company were not interested in establishing precedents. Even the granting of the long-awaited charter by the 1870 session of the legislature was considered of trifling importance.[80] Like most investors, they were impatient to realize some return on their money, and the Beef Slough Company, for all of its many accomplishments, managed to secure fewer logs in three seasons than the 50 million feet confidently contracted in that first season of 1868. Rather than profits to divide, there seemed no end of costs to share.

In the late spring of 1870 the editor of the *Alma Weekly Express* foresaw a bright future for the Beef Slough Company now that it was "established on a firm basis." But the actual situation was somewhat more despairing.[81] Indeed, the combined cost of the improvements, physical and political, had bankrupted the company. Francis Palms reportedly acknowledged that in addition to the $20,000 worth of stock he had purchased in the Beef Slough Company, he had personally loaned the organization "one hundred thousand dollars, more than fifty per cent of which will prove an utter loss."[82] Palms, for one, was willing to invest no more.

Ingram would have been understandably pleased had he known the extent of the losses suffered by the "outsiders" in their efforts to profit from speculation in Chippewa valley pine land. In a sense, the local operators had withstood the first assault, and if there had been some difficulties along the way, and even some

---

[79] Frederick K. Weyerhaeuser, *Trees and Men* (New York: Newcomen Society in North America, 1951) IV, p. 9.

[80] *Private & Local Laws,* 1870, Chapter 299.

[81] *Alma Weekly Express,* June 16, 1870.

[82] Randall, *Chippewa Valley,* 144; *Eau Claire Free Press,* January 28, 1875.

desperate moments, on the whole they seemed none the worse for the experience. Unfortunately, however, at least from the standpoint of the residents, the worst was yet to come. The success of Ingram and the other resident mill owners in the late 1860's had been achieved at the expense of those who, while they had considerable capital, had no special knowledge and experience in the lumber business. Palms and his associates were in no way committed to their effort in the valley of the Chippewa or to pine land there or elsewhere; they were speculators whose only real interest was a return on their investment. Soon Ingram and the other Chippewa millmen would be contending with parties who, if they had no more in the way of financial resources, were far more knowledgeable and persistent about their investments in lumbering.

The change in the opposing personnel was the result of a basic consideration. While men like Francis Palms could look elsewhere for opportunities to invest surplus funds, the sawmill owners along the Mississippi did not enjoy similar flexibility. They needed logs from the Chippewa, and in order to obtain those logs, the functions of the Beef Slough Company had to be continued regardless of what happened to the company itself. Consequently, on November 1, 1870, the entire operation—including booms, buildings, fixtures, and the all-important charter—was leased to Lorenzo Schricker, Elijah Swift, and Frederick Weyerhauser.[83] Although the contest for control of Beef Slough was far from ended, one of the new lessees soon demonstrated his confidence in the eventual outcome. That December Frederick Weyerhaeuser purchased 840 acres of Chippewa pine land from James Jenkins, a stockholder in the original Beef Slough Company.[84] Thus a new chapter opened in the history of the Chippewa valley, and in the history of lumbering. From that time forward, the future of Orrin Ingram would be much involved with the activities and ambitions of "the man," Frederick Weyerhaeuser.

Regardless of the uncertainties Ingram may have had concerning future developments, he nevertheless appreciated how much stronger his firm was in 1870 than it had been a decade

---

[83] Hidy, *et al.*, *Timber and Men*, 49–50; Matthew G. Norton, *The Mississippi River Logging Company* (n.p., 1912), 13–14.
[84] Hidy, *et al.*, *Timber and Men*, 50.

earlier. In the course of the past ten years he and his partner Kennedy had survived a fire, which destroyed the original sawmill, and a feud, which finally resulted in the dissolution of the original organization. But they soon began to experience the benefits of wartime inflation and prosperity. In the process, they were propelled into expanded production, and they pressured the available men and machinery to take utmost advantage of the highly favorable market conditions. Then, in the wake of postwar adjustment and deflation, they gave attention to the problems of marketing. This was especially important if Ingram & Kennedy was to continue to produce lumber at the high levels which peak demand had encouraged. Ingram was not one to reduce production simply because selling was no longer an easy task. He made new arrangements and accommodations, apparently without undue effort. He expanded retail facilities in southeastern Minnesota, and, in the process, formed a new partnership. Another partnership resulted from the establishment of a large wholesale operation at Dubuque, with supporting retail yards scattered among Iowa farming communities to the west. Consequently, much of the Ingram & Kennedy lumber was thereafter designated for yards associated with the Ingram & Kennedy organization.

Toward the end of the decade, nonresident speculators forced Ingram into what he considered to be a premature concern over pine land and ways of keeping his busy sawmills supplied with logs. The original group of "outsiders" sought to dispose of a portion of the annual pine harvest of the Chippewa valley to sawmill operators situated along the Mississippi. They were too impatient, however, to tolerate the large initial expense of such an effort and the accompanying delay in any return on their investment. Consequently, they sold out to the very Mississippi River millmen they had hoped to supply. The latter, led by Frederick Weyerhaeuser, had all of the patience necessary. Indeed, they had no real alternative to an eventual dependence upon the forest resources of the Chippewa basin. And so the lines of battle were redrawn. Orrin Ingram doubtless understood that the confrontation involved serious and far-reaching questions, but he also understood that if and when a compromise seemed in order, he and his firm were fairly well situated to strike a hard bargain.

CHAPTER V

Western Markets
and Chippewa Lumber

IN THE COURSE of the Civil War, manufacturers of iron, processors of leather, growers of grain, and many others—including producers of lumber—sought to increase their yields in order to take maximum advantage of the big demand and high prices. The focus of their efforts centered on their plants and the production phase of their operations. In many cases the same machines and fewer laborers were called upon to produce or process larger and larger amounts. Towards the end of the fighting, however, the northern economy began to anticipate the adjustments of a nation at peace. Demand diminished, and selling became a slow and difficult task at best. Consequently, the owners and managers of industry and agriculture were compelled to devote more of their attention to marketing, particularly if they planned no reduction in their output. Production remained important, of course, but no longer to the exclusion of other considerations.

Orrin Ingram fully appreciated the need for improvements in his firm's marketing operation. He had begun to extend and expand the retailing and wholesaling activities associated with Ingram & Kennedy immediately following the end of the war. But problems related to the marketing of lumber continued to command a large portion of his time and energy throughout the decade of the 1870's.

No successful lumberman was long able to involve himself in a single activity, such as logging, and ignore the problems of production or selling. It was necessary that he be aware of all of the steps involved in the business of making and marketing lumber, even if circumstances usually forced him to concentrate his efforts on a single phase in the process. With the passing seasons his priorities would change as one or another activity became more or less critical. Initially, for example, Ingram had concentrated capital and efforts on matters directly related to sawing logs into boards, timbers, shingles, and lath. But when these articles began to accumulate at the sawmill or at the yards associated with the sawmill, he assumed new responsibilities, attempting to locate buyers for his lumber product. In later years, still another orientation would be required as it became apparent that the supply of pine for the sawmills could no longer be taken for granted.

In the immediate postwar period, when Ingram began to evidence an interest in enlarging the marketing network of Ingram & Kennedy, he was considering adjustment common to most northern producers and processors. Still, there were some special problems which concerned the lumber industry; and, indeed, some problems were unique to the lumber industry of the Chippewa valley.

The fundamental question had to do with the location of the market. Through much of the last half of the nineteenth century, the great and ever-growing market for pine lumber very nearly coincided with the advancing farming frontier of the trans-Mississippi west. It was possible to follow the progress of that frontier as closely at the desk of a clerk in a wholesale lumber yard as anywhere else. The delineation of the boundaries of "natural" or "legitimate" markets for various distribution centers is interesting. These territories were, for the most part, situated adjacent to and directly west of the wholesale yard. Accordingly, when Major Day spoke of his market, he was referring to the Iowa towns along the railroads leading west out of Dubuque. Quite naturally he did not appreciate invasions of his territory, although he was enough of a businessman to realize that on occasion conditions required that efforts be made to sell in districts other than one's own. On more than one occasion he described the nature of these adjustments, observing that since "Our legitimate customers in Iowa are not wanting much lumber . . . we are obliged to skirmish round for trade."[1] Grad-

---

[1] Day to Ingram, May 15, 1884.

ually, however, as more and more railroads intersected the established east-west routes, marketing areas became generally more diffused and increasingly competitive.

Major Day's description of what he considered to be his "legitimate" lumber market was only another reflection of the basic east-west orientation of migration, communications, and commerce. Although the relief of the United States generally follows north-and-south lines, this geographic configuration has had relatively small effect on the direction of the movement of people and products. Allowing for local exceptions, the waves of settlement progressed westward nearly at right angles to the lay of the land, and the roads and rails which tied the expanding nation together assumed a similar east-west orientation.[2] When people or goods moved north or south, more than likely it was merely a side-step, a temporary realignment in order to avoid a particular obstacle. Once placed on a more favorable tack, people and goods moved west again, almost as if by compass bearing.

Lumber produced in the Chippewa valley exemplified this temporary exception to the east-west pattern of commerce. Initially, and for many years to follow, lumber rafts moved out of the Chippewa to river ports along the Mississippi. These rafting operations continued long after the arrival of the railroad. There were a number of reasons why many of the millmen of the valley, including Ingram, rafted their lumber downriver even after the option of rail transportation had been provided.

The most obvious advantage of water over rail transportation was the comparative cost of moving such a bulky and relatively inexpensive item. But the consideration of costs does not explain why producers elsewhere shipped by rail instead of water, making the change as soon as the railroads could provide the service and the equipment that was necessary. Clearly one of the reasons that Chippewa operators continued to raft their lumber had to do with geography, specifically the location of the Chippewa in relation to other streams flowing out of the pine district. In effect, the Chippewa valley millmen were flanked on either side by other producers. To the east at Wausau, Stevens Point, Grand Rapids (now Wisconsin Rapids), and other Wisconsin River

---

[2] For an interesting discussion of this subject, see Edward Louis Ullman, *American Commodity Flow* (Seattle: University of Washington Press, 1957).

towns, lumber was being sawed in considerable quantities. To the west along the St. Croix River, production was increasing rapidly, especially in and around Stillwater, Minnesota. Somewhat farther to the west were the extensive and busy sawmilling facilities of Minneapolis.

Obviously a direct east or west movement of lumber out of the Chippewa valley would encounter immediate competition. With the construction of the Wisconsin Central and Wisconsin Valley railroads in the early 1870's, the Wisconsin River lumbermen had begun to take an increasing interest in the more local markets of central and southeastern Wisconsin.[3] Had the Chippewa producers made an effort to ship their lumber east and southeast towards Milwaukee and Chicago, they would likely have found themselves in a disadvantageous position in the resulting competition with their Wisconsin valley counterparts. Somewhat the same situation confronted them to the west. Although the Dakotas and Manitoba were attracting large numbers of immigrants, Minnesota millmen enjoyed the advantages of proximity in supplying those areas with lumber. While none of these markets was exclusively the domain of a single sawmilling center or group of suppliers, relative advantages and disadvantages had to be considered.

Consequently, for many seasons the lumber from the Chippewa made what amounted to a strategic movement by water down the Mississippi. It does not follow, however, that the actual consumers of the lumber product were situated along the river banks. If that were true, one might expect a somewhat equal distribution of towns on either bank visited by the immense rafts. But the river ports where the rafts tied up were largely on the west bank, because they were merely the points of transshipment. They were the convenient places where the lumber was pulled from the water, cleaned, planed, and seasoned, before being loaded on cars and resuming the westward journey by rail. Water was therefore no substitute for rail transport; it was simply an inexpensive means of moving to a more favorable point from which to begin the journey by rail.[4] Lumber from the Chippewa valley had to move with advantage towards the western markets.

---

[3] W. H. Glover, "Lumber Rafting on the Wisconsin River," in the *Wisconsin Magazine of History*, [Part One] 25 (December, 1941); [Part Two] 25 (March, 1942).

[4] The relatively low cost of rafting lumber was, of course, a most important

Ingram's efforts to improve and expand the marketing arrangements of Ingram & Kennedy did not always proceed smoothly. One of the larger complications which he faced in the decade of the 1870's was another depression, the most severe since 1857. As had been the case with the earlier one, however, the contraction had surprisingly small effect on the successful operation of the Ingram & Kennedy organization or even on its plans and programs for expansion. Prices in the lumber markets of the west had not been high before the failure of Jay Cooke & Company in September, 1873, and they were not to be significantly lower afterwards. The apparent reason for this relative stability was that farm population increased at a rate of 25 per cent during the 1869 to 1879, while both the number of farms and the acreage of land in use increased 50 per cent.[5] Under these circumstances, customers for lumber were not lacking; if prices were not high, neither were costs, and profits remained substantial.

The decade began inauspiciously for lumber producers and dealers. The season of 1869 had not been a particularly good one for Ingram & Kennedy. In mid-August, usually one of the best periods for selling, wholesale prices on common lumber at Dubuque had fallen to $12, and in September the markets were further depressed by the onset of what seemed a general monsoon. Heavy, continual rains made the delivery of river lumber on high water difficult at most places and impossible at some. If river transportation was complicated by the rising Mississippi, movement on land nearly ceased altogether as mud kept customers at home. Major Day reported that in the vicinity of Dubuque, rains fell on a schedule "just sufficient to keep the roads

---

consideration. Had the costs of water and rail movement been at all comparable, rafting operations would have ceased immediately following the arrival of the first train at the site of production. But the costs varied greatly. Figuring an average freight rate of fifteen cents per hundred pounds from Eau Claire to Hannibal, Missouri (and these rail rates changed significantly from season to season, and sometimes from week to week), and the weight of the dry white pine lumber at an average of 2500 pounds per thousand board feet, then the cost per thousand feet by rail would be approximately $2.75. This would exceed the average cost of water transport by at least $1 per thousand feet. In a series of articles entitled "Navigation on the Chippewa River in Wisconsin," Captain Fred A. Bill has estimated the cost of the entire trip from pinery to Mississippi markets by water to average $1 per ton of lumber, and his estimate would seem to be fairly accurate. See *Burlington Post*, December 13, 1930.

[5] O. V. Wells, "The Depression of 1873–79, Historical Aspects of Agricultural Adjustment," in *Agricultural History*, 2 (July, 1937), 240–243.

impassable or nearly so," and Archer described the roads in south-eastern Minnesota as simply "horrible." He watched the Missis-sippi rise to the point where some of the lumber piles in his Min-neiska yard began to float about, hastening to assure the Eau Claire office that everything was "perfectly safe," if muddy.[6]

Conditions were no better away from the river, where farmers impatiently awaited an opportunity to thresh their wheat. To make matters worse, Chicago lumber was riding in by rail on even more favorable rates than had previously been true. Playter, quoting a Dubuque dealer who had just returned from a trip through western Iowa, described that section as being "all cov-ered with water, but where there was a dry spot it was covered with *Chicago lumber.*"[7]

Receiving almost no funds at the Eau Claire office, Ingram dis-cussed the possibility of shutting down the large mill before the end of September in order to save on wages, since they seemed unable to meet the costs of production. But although business at the Dubuque yard had been poor, Major Day found the sugges-tion of an early end to sawing an unacceptable solution to their fiscal problems. He immediately forwarded his views on the subject to Ingram, at the same time advising him that they would sell from the water what they could out of the next raft to arrive at Dubuque "to enable you to realize some cash from it."[8]

The Major's chief objection to closing down the large mill at Eau Claire had to do with the assortment of lumber then on hand at the Dubuque yard. He considered his inventory insuffi-cient to meet the demands of winter trade, particularly the amount of dimension lumber, most of which was produced by the large mill. In his letter to Ingram in which he listed the various yard shortages, Day explained that "unless we receive some help from you in this particular [we] will be obliged to make large cash outlays to the mills here next spring & patronize competing yards largely this winter." In other words, the Major argued that any decision to stop producing lumber earlier than necessary would likely result in the briefest of benefits, and the old problem of a cash shortage would only be aggravated in the process.

---

[6] Archer to Ingram & Kennedy, September 24, 1869.
[7] Playter to Ingram & Kennedy, September 19, 1869.
[8] Day to Ingram, September 9, 1869.

Day thought that he needed more lumber at Dubuque. Ingram knew that he needed more cash at Eau Claire. But Ingram did not seriously think of closing the large mill as he had threatened. Rather he was seeking to encourage Day and others to make the maximum effort to forward cash to Eau Claire. Although the firm had never been in a healthier state, Ingram found himself suffering through another stringent period when he was severely pressed to meet payrolls and due bills. Yet as unpleasant as it may have been to forage about for operating funds, cash uninvested was so objectionable that they would continue to pass from crisis to crisis, with ever-increasing assets and barely enough liquidity to manage.

Despite the troubles which had raised the possibility of an early shutdown, the sawing and delivery of Ingram & Kennedy lumber continued in the fall of 1869 until the arrival of an early winter halted activity at the mills and along the rivers. By November 20, Dubuque was blanketed under eight inches of snow, surprising the Major with a considerable amount of lumber—some thirty cribs—still in the water. Day's optimism, however, enabled him to look beyond the immediate prospect, which remained rather gloomy. As he wrote to Eau Claire in late November, "unless we start other yards, we will be obliged to quietly await the late winter & early spring trade."[9] Since he found idleness intolerable and was unable "to quietly await" anything, it was not surprising that the Major proceeded to arrange for the purchase of a retail yard at Independence, Iowa. He would be the first to agree that when lumber sold for good margins in the water, it might just as well be sold there. At other times, alternatives had to be considered, and the most obvious was to assume the responsibilities of retailing your own lumber. Indeed, the Major seemed to operate on the premise that when the retail yards were not purchasing lumber, it was an ideal time to purchase retail yards. Sales or not, Ingram, Kennedy & Day did business, even if forced to make their own.

There was plenty of water in the spring of 1870, and experienced woodsmen claimed never to have participated in cleaner drives on the streams of the Chippewa valley. Although the Mississippi rose to dangerous levels at many points, it receded before doing

[9] Day to Ingram & Kennedy, November 20, 1869.

any serious damage to the yards in Minnesota or at Dubuque. As the river fell, trade along the line increased. Only a freak accident involving Major Day's wife, the effects of which kept her an invalid until her death in the fall of 1871, detracted from the encouraging prospects for business. Ingram was genuinely sympathetic, and as helpful as circumstances would permit, but the Major's personal tragedy could not lessen the dependence of the Eau Claire office on Dubuque for funds. The business there had received unexpected assistance when the sawmill of a major competitor, the Dubuque Lumber Company, burned down in early April. While the well-mannered Major Day declined to comment on the possible advantages resulting from the adversity of his neighbor, Playter was less reticent. He observed: "Moore's mill not being in operation is going to make quite a difference with [the] market here, and if Lumber is kept at a price, so there will not be any inducements for dealers to go to Chicago, [I] think there will be more Lumber handled here this Season, than any previous one."[10]

Playter, who had been sent to Dubuque to assist the Major following Mrs. Day's accident, found the yard in excellent shape and with no lack of customers. As the season got underway, he became increasingly optimistic about the Dubuque operation and wrote to friend George Potter, then keeping books in the Eau Claire office, that "the prospect for Lumber is better than I expected." In addition to the large trade at the main yard, he predicted that Ingram, Kennedy & Day's retail yards at Winthrop, Independence, Pomeroy and Janesville would be demanding "considerable Lumber." A week later he thought that prospects looked "still better" and saw no reason why Ingram & Kennedy lumber would not "take the lead in this market this year."[11] Playter, of course, was prejudiced.

But even during the good business seasons, a lack of ready cash was not an unfamiliar problem. Although as Playter predicted, the volume of 1870 sales was large, somehow the return was only a fraction of what had been expected. Ingram, who was anxious to purchase some pine land, expressed his unhappiness on numerous occasions over his continued lack of funds. As the summer

---

[10] Playter to Ingram & Kennedy, May 19, 1870.
[11] Playter to Potter, June 1 and June 8, 1870.

wore on and the cash situation failed to improve, Ingram addressed some unusually strong remarks in the direction of Dubuque, demanding from Major Day some explanation for his obvious inability to forward receipts at any reasonable rate. Day responded tactfully, acknowledging that "from the magnitude of your investment here you have reason to expect liberal assistance . . . when you require it." But the problem then experienced in Eau Claire had more to do with the times than with the policies or personnel of Ingram, Kennedy & Day, and the Major proceeded to describe the efforts they had made to collect and the lack of success they had enjoyed in these efforts. The theme was old and familiar: the trouble was not in selling but in getting the money. Ingram, Kennedy & Day had more than $16,500 in outstanding accounts among its country customers on August 1, 1870, with an additional $2000 owed by parties in Dubuque. Also, the company's various retail yards were indebted to the main yard by some $27,000, making a total in excess of $45,000 on the books but currently not available to Ingram.[12]

As was the case at Dubuque, business at Wabasha and Minneiska seemed more active than the prices might indicate, common boards selling for $13.50 per thousand feet in June. Generally receipts tended to be better at the Minnesota yards, although Albert Archer expressed some dissatisfaction with his $1,500 total for May, maintaining, as Randall had often maintained, that he could easily have exceeded $2,000 "if I had the lumber." But in spite of shortages, first in fencing and later in shingles, customers continued to visit the Minneiska yard in record numbers. By the middle of June, Archer reported his first thousand-dollar week. Furthermore, there was no indication that demand would soon diminish, the Minneiska manager observing that "a good army are talking about building barns. . . ."[13]

The week that Archer collected $1000 at Minneiska, Gill enjoyed comparable success at Wabasha, receiving more than $1200 and making similar noises about his insufficient inventory. In the main, however, the problems experienced in the spring and early summer of 1870 were of a pleasant variety, having more to

[12] Day to Ingram, August 6, 1870.
[13] Archer to Ingram & Kennedy, June 3 and June 12, 1870.

do with a large demand than any sudden drop in supply. Even the expected decline before harvest did not materialize as a result, according to Archer, of "the war between France and Prussia [which] has run wheat up in two days from 97 cts. to $1.17."[14] If the demand for lumber seemed to be exceeding the supply at many points, the price for "Chipeway lumber" in the water nevertheless held stubbornly close to $14 throughout most of the season.

With the volume of sales relatively large and prices relatively low, pressures increased not only to exceed previous levels of production but at the same time to reduce costs. In the process, concern over quality diminished. Logs previously rejected moved up the bullslides into the mills, where men and machines worked beyond limits of efficiency; and the lumber produced was culled with less care as competition became less severe in the downriver markets.

Individual dealers, especially those ordering large lots, could request so much of this kind or that length and none of the other, but such requests were seldom taken too seriously, at least not at the sawmills. If the sawyers could satisfy a customer's general directions for sawing and at the same time cut the log to the best possible advantage, they would do so—but only because the stated preferences happened to coincide with the characteristics of the log. In other words, it was the sawlog which determined the kind and quality of lumber produced, and a mill owner who operated on any other policy did so with no chance for consistent margins.

Nevertheless, some customers continued to insist, through good times and bad, on their rights to demand special treatment at regular prices. D. T. Robinson of Rock Island, Illinois, who referred to himself as "the grumbler," often annoyed the agents and mill superintendents of Ingram & Kennedy with the detailed sawing instructions he invariably attached to his orders for lumber. He must have realized that such instructions were unlikely to be followed, at least not unless a firm was in dire need of business. The response of Ingram & Kennedy to such directions assumed nearly a standard form: they would be pleased to fill the bill, as nearly "as the Logs would make &c."[15] All par-

---

[14] Archer to Ingram & Kennedy, July 18, 1870.
[15] Potter to Ingram & Kennedy, July 14, 1870.

ties recognized that this was simply another and more diplomatic way of saying, "We will deliver our mill-run lumber to the amount requested."

The only reasonably accurate method of predicting quality of the lumber was by knowing the kind of logs from which it was to be sawed. Therefore, if Ingram wanted to provide a customer with a particular quality, he could suggest that they saw out that order at a time when the logs coming into the mill most closely matched the customer's preference for lumber. This might require, however, that the customer wait a longer time for delivery, and, more often than not, he would not stand the delay.

Not uncommonly, those customers who ordered with such great detail were prone to complain about the lumber subsequently delivered them. When there was a question about quality, or the grading, or the kinds of lumber received, an agent from the company would normally investigate and, if the complaint proved to be justified, an adjustment in price would be allowed. Despite the tendencies of some customers, adjustments in price did not automatically follow the complaint. For example, during the summer of 1871, Playter made a special trip to Rock Island to look over some of the lumber recently received by Robinson which had been laid aside as proof of poor sawing, "the grumbler" insisting that some reduction in price ought to be made. In apt illustration of the current attitudes toward quality of lumber, agent Eb reported that he had seen in Robinson's yard "a good many of the Long Joists [that] were unevenly sawed, Some were 3 inches thick at one end & [at the] other 1½ Inch, some 3 x 4 & 4 x 4 varied a good deal (from 2½ to 5 Inches) also some of the Timber was badly sawn, and among the Marston Mill Lbr. the Boards were a little wavy. . . ." Playter was unimpressed with the evidence offered by Robinson, reporting to Eau Claire that he failed to "find anything so awfull bad about it."[16]

Business held to a high volume throughout the 1870 season, and by October the financial situation of Ingram, Kennedy & Day reflected considerable progress. On October 6, the Major sent better news from Dubuque than had previously been possible, informing the Eau Claire office, "Paid your draft for $4,000

[16] Playter to Ingram & Kennedy, August 22, 1871.

today, owe the bank nothing, & have about $2,000 to our cr[edit]."
The bank balance, of course, would not long be allowed to remain. Ingram, with some continued reluctance, was busy buying timber land and stumpage,[17] content to read the newspaper
accounts of the marriage of twenty-year-old Sara G. Thorp,
friend Joseph's daughter, to the sixty-four-year-old "prince of
musical artists," Ole Bull, in what was truly the social event of
the season in Madison, Wisconsin.[18]

Aside from the flooding experienced along the Chippewa in
the early fall of 1870, these were ideal days for those engaged in
the lumber business. Money was again moving freely. Major
Day informed the Eau Claire office that he would be able to
provide between $15,000 and $18,000 during November, noting
that his trade at Dubuque continued "quite brisk from the
yard."[19] Archer was also keeping busy at Minneiska where, in
spite of a seventy-four-cent bushel price for wheat, he counted
up another thousand-dollar week in early November. Comple-
menting the returns from their own yards were the orders from
the many wholesale customers, who were now forced to compete
for Ingram & Kennedy lumber. Even the chronic complainer
D. T. Robinson decided to make his 1871 needs known early,
addressing a request in October for a million feet to be delivered
the following spring, on which he was willing to advance $6000
before the first of January.[20]

With the situation looking rather comfortable in the Eau
Claire office, Ingram concluded to take another trip east, spending most of his time in the Ottawa River valley, where he had begun to learn the lumber business not so long ago. During the
1870 visit he wrote often to George Potter, his Eau Claire bookkeeper, describing the many improvements that had recently

[17] The distinction between timber land and timber stumpage is an important
one. In both cases the trees are the object of purchase, but in the case of
stumpage, it is only the trees that are purchased; the land remains in possession
of the stumpage seller. It is exactly the same procedure common in some areas
as concerns hay. Farmers may purchase hay stumpage—that is, the privilege of
harvesting the hay crop—and the ownership of the land is unaffected. Naturally
most lumbermen preferred stumpage purchases since the costs measured in board
feet would be less and the problems of taxation and land disposal someone else's.
[18] Eau Claire Free Press, September 29, 1870.
[19] Day to Ingram & Kennedy, October 26, 1870.
[20] Robinson to Ingram & Kennedy, October 27, 1870.

been made in the city of Ottawa. In spite of these improvements, Ingram decided "it still has a Canada look." There was something missing in Ottawa. The Canadian city could hardly have been more provincial than Eau Claire, but he did not see or feel the same spirit of optimism along the banks of the Ottawa that he did along the banks of the Chippewa. Nevertheless, he was much impressed by the sawmills he visited and the lumber he inspected, reporting that twelve-foot boards then wholesaling in Ottawa at $7 per thousand feet would bring $16 in the Dubuque market. Clearly, the comparative advantages for making money belonged to the producers of the west. Ingram thought there was probably "more money invested in one Mill here than [in] all [of] the Mills in Eau Claire, but they cannot saw lumber in proportion to the investment." Or, reversing the example, with an equal investment in Eau Claire, they could make "double the money in lumber."[21]

Yet by some manner or means, the sawmill operators of the Ottawa valley were obviously, almost ostentatiously from the standpoint of their facilities, enjoying success; and Ingram was not so blinded by local pride that he failed to note that his former employers and other lumbermen of the area were doing very well. Always alert to new opportunities, he reflected on the sources of profit for the product of the Canadian mills and concluded that something had to be said for the English markets. Indeed, five years later in the fall of 1876, Ingram would still be wondering about the feasibility of shipping deals from Eau Claire to England via New Orleans. In some respects, such a possibility seemed not unreasonable, since the movement to New Orleans would be entirely by water, like the movement to Quebec down the St. Lawrence. Furthermore there seemed no apparent advantage in Quebec over New Orleans as the point of transshipment, since Quebec was frozen in for so much of the year. Ingram envisioned floating the deals down to New Orleans during the rafting season in time for a winter voyage to England. He knew that many of the ships involved in the Quebec-England summer trade found employment in the cotton and pitch-pine trade from ports like Savannah, Mobile, Pensacola, and New Orleans during the months that ice closed the St. Lawrence.[22]

[21] Ingram to Potter, December 7, 1870.
[22] James Cunningham (Ottawa, Ontario) to Ingram, November 11, 1876.

Chippewa deals would therefore be just another product which could be handled within a well-established commercial arrangement.

Even assuming, however, that the three-inch planks could reach New Orleans unwarped and unstained by their lengthy journey in the waters of the Mississippi, which was no small supposition, other factors interposed. To begin with, the lumbermen of the Chippewa valley required no distant option to aid in the disposal of their pine product. Although there were occasional periods of weakness, for the most part they could not keep abreast of the growing demand, and any diversion of a portion of their lumber from the real markets was distracting to buyer and seller alike. Perhaps the major drawback which discouraged shipment of lumber to the established European markets was that any such involvement on a large scale would consume most of the higher quality logs, thus making the sale of the average and lower qualities of lumber all the more difficult. Consequently, although Ingram and others may occasionally have considered a variety of marketing alternatives, they soon returned to the business at hand: making cheap boards and timbers for their farmer customers west of the Mississippi.

During the majority of seasons, Ingram was satisfied if his firm kept its yards and regular customers supplied according to their needs. This was no small task. Each succeeding season seemed to require lumber in larger amounts. But as the level of production of the Ingram & Kennedy mills continued to increase, Ingram personally spent fewer hours in the plants. He left the problems associated with production largely in the hands of Kennedy and the individual superintendents, dividing his own time between matters relating to the supply of pine and the sale of lumber.

By the 1870's, Ingram had begun to make purchases of pine land, although he entered into this activity with reluctance. At the same time, he was also giving continued consideration to the development of new and larger marketing arrangements. In effect, Ingram & Kennedy was gradually becoming an integrated business organization, but only in part as a result of conscious effort of the partners. To a large extent, circumstances were pushing Ingram & Kennedy into expanding their activities. It was no longer possible to remain strictly a producer of pine lumber.

Upon returning from his Canadian trip in the winter of 1870–1871, Ingram turned his attention to a consideration of a market area whose potential he had long appreciated: Hannibal, Missouri. By then it had become a common practice for Chippewa River millmen to make arrangements with wholesale dealers at Mississippi River towns, often to the point of forming partnerships, as Ingram and Kennedy had done with Major Day at Dubuque. Now Ingram was looking toward the establishment of a similar association with a Hannibal lumberman, but no Major Day was initially evident.[23] In any event, Ingram & Kennedy lumber began to arrive at Hannibal in quantity during the 1871 season.

These deliveries, expedited by the company boat *Clyde,* did result in an increased proportion of cash sales. The benefits, however, were somewhat offset by the appearance of additional problems. In the first place, although the Eau Claire mills of Ingram & Kennedy were producing more and more lumber, the firm still had to be selective in its arrangements for disposition of the product. With Ingram convinced in 1871 that some lumber must be sent to Hannibal, it soon became apparent that there would be real difficulty in keeping the established yards and old customers adequately supplied. A situation developed somewhat reminiscent of years earlier when Kennedy had observed that partner Dole had been spreading a little lumber over a large area. As Davis, Bockee & Garth, J. J. Cruikshank, Dulany & Mc-Veigh, Hearne, Herriman & Company and D. Dubach & Company pulled Ingram & Kennedy cribs out on the levees of Hannibal, manager Gill in Wabasha was urgently requesting delivery of a hundred thousand feet for himself and a similar amount for Archer at Minneiska, "as soon as possible." Meanwhile, Major Day was complaining from his desk in Dubuque, "Boards will be the burden of my prayer until we get some in the yard."[24]

Yet in spite of the diversion of lumber to Hannibal, Ingram, Kennedy & Day were selling increasing amounts at Dubuque. The Major reported that during the first six months of 1871 they had sold 4,250,000 feet, three-quarters of a million more than in

---

[23] N. C. Amsden to Ingram, July 2, 1871.
[24] Gill to Ingram & Kennedy, July 14, 1871; Day to Ingram & Kennedy, July 13, 1871.

the same period of the preceding season. In addition, he estimated that the Dubuque yard should be provided with at least 5
million more feet before the end of rafting operations, a requirement made the more difficult by Ingram's insistence that substantial quantities be sent downriver to Hannibal.

As a further initial complication, Ingram & Kennedy lumber
did not seem to command prices in the Hannibal market which
warranted the additional effort and expense involved in rafting
the greater distance. John C. Daniels, a commission agent from
Keokuk who worked that section of the river, commented on
this problem when explaining the arrangements he had made for
a sale at $15.25, including his 2½ per cent commission charge.
Daniels informed Ingram and Kennedy that they had to expect
low prices for a time at least, because "no one new [sic] your
lumber."[25]

Such figures, when balanced against the additional expense
per thousand feet of running from Dubuque on down to Hannibal, were not easily justified. Yet the price negotiated by agent
Daniels with Davis, Bockee & Garth proved to be one of the
better trades of the summer; and when Playter had to sell an
August raft at the Hannibal levee for $15 per thousand, he ad-
vised the Eau Claire office that, for obvious reasons, he preferred
that Major Day did not learn the details of the trade.[26] Although
Ingram appreciated the difficulties encountered by Daniels and
Playter in selling to advantage, he also knew that there was only
one way for Ingram & Kennedy lumber to achieve a reputation in
the Hannibal market: they had to endure an initial period of
introduction, even if it proved necessary to sell at an absolute
loss. Such a sacrifice, however, was not required.

A further difficulty of varying extent in the delivery of lumber
to Hannibal involved the additional distance and time of such
a trip, and the condition of the river. Aside from the numerous
islands, sand bars, and bridges, only a few serious problems confronted the pilot during most stages of water in the raft journey
from Reads Landing to Dubuque. But from Dubuque downriver
to Hannibal, two sets of rapids intervened which caused frequent
delay and occasional damage and destruction.

[25] Daniels to Ingram & Kennedy, June 5, 1871.
[26] Playter to Ingram & Kennedy, August 27, 1871.

At the head of the upper or Rock Island Rapids was LeClaire, Iowa. The usual practice was to stop the lumber raft at that river port in order to take aboard a special rapids pilot, who would then take charge of the boat and raft, directing them down through the dangerous stretch to Davenport, where he went ashore. These upper rapids were almost never attempted at night, and the delay at LeClaire was often considerable as rafts bound for points below waited in line for a clear channel and a rapids pilot. Even more serious an obstacle were the lower rapids, located between Montrose and Keokuk. At Montrose, as at Le-Claire, local pilots were hired to guide the rafts through the rough water.[27]

During 1871 the Mississippi fell to extremely low levels by early June, causing unseasonable difficulty and delay in negotiating the lower rapids. Many large sandbars were exposed between Montrose and Keokuk, one of which the *Clyde* encountered on her first trip to Hannibal. But as crews on the Ingram & Kennedy rafts soon learned, in the river below LeClaire "rubbing sand" was no rare occurrence, and they would come to consider themselves fortunate if delays were brief and damage slight in the course of their journeys to Hannibal.

Rafting proceeded in the summer of 1871 notwithstanding these unfavorable conditions. By August there was not enough water on the lower rapids to chance taking the *Clyde* below Montrose. So she was tied up and, with the thermometer standing at 105° and "not a breath of air," the raft was floated the rest of the way down to Hannibal. Playter, observing how badly "the men suffered on [the] Raft," could not help but wonder aloud whether all of the discomfort and difficulty in getting the lumber to Hannibal was worth it, especially when one considered the prices at which they were forced to sell.

But Ingram was not just selling lumber in 1871. He was building for the future, creating a demand for Ingram & Kennedy lumber which he thought would more than compensate for any hardships experienced in the process. Just what convinced him that Hannibal was the point to make such an effort is not altogether clear. Perhaps it was more than coincidence that none of the

[27] For an interesting description of the river as it was, see Mildred L. Hartsough, *From Canoe to Steel Barge on the Upper Mississippi* (Minneapolis: University of Minnesota Press, 1934), 76–89.

member firms of the newly-formed Mississippi River Logging
Company were from Hannibal.[28] Or possibly the early predictions
of Whitehill, and later of Alexander Dole, regarding the promise and the future of that area, had not been forgotten. Whatever the reason, by mid-summer Ingram appeared even more confident that Hannibal would soon prove to be a most profitable
market for Chippewa lumber. The only consideration which
prevented the opening of a wholesale yard there was the lack
of someone with some experience and some funds who was willing to take a part interest in the enterprise.[29]

For all of the promise and potential of the Hannibal market,
however, considerations had to be given the problems which resulted from rafting such a distance. Every difficulty and delay
experienced in the rafting operation increased expenses and reduced profits. Aside from the natural hazards of rapids and rocks,
islands, bends and bars, the pilot had to maneuver his immense
wooden floe past an increasing number of man-made obstacles
—piers, dams, bridges—few of which had been constructed with
the problems of raft navigation foremost in mind.

The lumbermen who bore the costs of water shipment were
constantly reminded of the special importance of the prudent
handling of their rafts. The value of a good and responsible
pilot could hardly be overestimated, and it was not only Mark
Twain who marveled at the knowledge and ability of those, like
himself, who worked the Mississippi. Joseph Buisson, a pilot for
Ingram & Kennedy in the early 1870's, said that in order to learn

[28] The Mississippi River Logging Company, successor to the Beef Slough Company, was organized in December, 1870, by nonresident sawmilling interests dependent on Chippewa pine for their operation. Its motivating force was Frederick Weyerhaeuser, but included in the organization were the major producers
along the Mississippi from Winona, Minnesota, southward as far as St. Louis.
The reason for their association was clear enough: the common need for logs
and the assumption that by combining their resources they could accomplish what
none could do singly—defeat the local millmen who opposed any exodus of
sawlogs from the Chippewa valley.

[29] It should be noted that Ingram & Kennedy was not the first Chippewa valley
firm to seek such an arrangement in Hannibal, although the majority of dealers
there were better acquainted with lumber from sawmills along the Wisconsin
River. During the previous year, 1870, Porter & Moon of Eau Claire formed an
association with Sumner T. McKnight of Hannibal, exchanging one-third interests
in their respective operations. The organization which resulted, the Northwestern Lumber Company, became one of the largest and best known in the Mississippi River valley. See *Eau Claire Free Press*, February 17, 1870.

the river and the art of handling boats it required at least "five years of hard study with one who will exemplify the business in every detail." And such an opportunity to learn did not alone insure the development of a first pilot. Of equal importance, at least according to Buisson, was the possession of "nearve, good judgement, excellent ability and, lastly but not least, sobriety."[30] He may have been overly generous in this assessment, but it was true that good pilots were at a premium, and that some who called themselves first pilots might better have followed another profession. In any case, during the 1871 rafting season Joseph Buisson's skills were severely tested.

Low water, fog, and head winds continued to complicate the rafting operations as the season lengthened into September. By the middle of that month, the stage of the Mississippi prevented the *Clyde* from attempting to run over either stretch of rapids. Consequently, the rafts were compelled to continue their journey below LeClaire without benefit of the boat. Manned by a double crew on the oars, they floated downstream with difficulty. One raft went briefly aground on the upper rapids, although only two cribs were damaged seriously. Still, there was considerable delay and Playter reported to Eau Claire, "If we keep on you need not look for much money out of this Raft, as it will take what we get to pay expenses."[31]

As a result of the adverse conditions on the Mississippi, Playter was not at all confident about the prospects of getting over the lower rapids. He therefore landed the 112-crib raft at Burlington, Iowa, where he hoped to arrange a sale at $14.50 for the dimension lumber and $15 for the boards and strips. In support of this decision, he informed the Eau Claire office of a conversation with Eugene Shaw, then acting as the downriver agent for the Daniel Shaw Lumber Company. Shaw had reported costs amounting to fifty cents per thousand feet in order to run the lower rapids. Although that seemed an exorbitant charge, Playter doubted that it was any exaggeration and, for the benefit of his distant employers, he described the situation below Montrose:[32]

[30] Joseph Buisson Papers, State Historical Society of Wisconsin, Manuscripts Library, personal notes, no date.

[31] Playter to Ingram & Kennedy, September 17, 1871.

[32] Playter to Ingram & Kennedy, September 19, 1871.

They are now running over 4 Strings 10 long at a time. 16 Men
on each piece besides pilots, $15.00 a trip for Pilot & $2.00 for
use of Anchors placed on Rapids, so it would take us 3 days to
get 8 strings over, unless we hired men at Montrose at $3.50 for
a trip & three pilots, which would not be as well as to run over
with our own men, so that if we did not have to lay up for wind
at all, cannot figure expenses less than 40 cts. per M and should
we have any bad luck do not know how much it would cost.

On learning that agent Playter was attempting to dispose of
the raft above, Dulany & McVeigh of Hannibal addressed a
letter of some urgency to the Eau Claire office of Ingram &
Kennedy. Dulany & McVeigh wanted the lumber badly and
wanted Ingram to order Playter to bring the raft on down to
Hannibal regardless of how little water ran over the lower rapids.
This was the very expression of interest that Ingram had been
waiting to receive from Hannibal. Dulany & McVeigh even
proposed an increase in the sale price for the raft, raising the
figure by fifty cents to $16 per thousand feet. Of itself, such
an advance was hardly enough to convince Playter to resume
the journey, but since he found no customers at Burlington he
could only continue downstream to Hannibal. As expected, it
proved to be a tiring, trying and expensive five days, and the
$12,000 which was received in cash payment seemed an insuffi-
cient consolation. For all of the hardships experienced by Playter and his crew, however, Ingram felt encouraged. Hannibal
customers were clearly becoming interested in the lumber produced by Ingram & Kennedy.

But the troubles had not yet come to an end. On her journey
back upstream, the iron-clad *Clyde* struck a snag and sank just
above Burlington. Raising her was more difficult than initially
estimated, and Kennedy finally had to go down from Eau Claire
to supervise the efforts at recovery. By October 13, two weeks
after the accident, the raft boat was listing her way upstream again,
Kennedy hoping they would be able to reach Dubuque where
permanent repairs could be made. He was none too pleased with
the rescue efforts, remarking, "if she had been a wooden boat
they would have tore her all to pieces."[33]

It was the end of October before the *Clyde* was back in operation, and according to Playter's estimate, which included time

[33] Kennedy to Ingram, October 13, 1871.

lost in rafting, the total cost amounted to about $3000, making it "Rather an expensive accident."[34] With the raft boat having missed exactly a month of running, all energies were directed to make up some of the lost time. But it was already late in the season and there was snow in the air.

While the *Clyde* was being recovered and repaired, Joseph Buisson went on up to Reads Landing to float a raft to Dubuque. Even in that upper section of the river, conditions had become extremely unfavorable for running. Playter told of some rafts passing Dubuque in mid-October that had left Reads seventeen days earlier, a trip that would normally require three to four days. Because the river continued to fall, the supply of lumber in the Mississippi markets also diminished, which served only to increase the frustration at not being able to move lumber downstream with greater dispatch. As usual, however, the situation could have been worse. For the most part, Chippewa lumbermen were still able to get their rafts out of the tributary and into the Mississippi. All were not so fortunate, Playter reporting in late October, "From what I hear . . . there will not be water enough to bring out any lumber from the Wisconsin so what is there will have to stay until Spring."[35]

Rainmakers must have been in great demand in the late summer and fall of 1871. The land grew parched, the streams dried up, and the rivers ran shallow. Even the lumbermen became relatively less concerned about grounded rafts than about the dangers of fire, but none could foresee the extent of the tragedies which would result. On October 8, fire consumed some 18,000 buildings in the heart of Chicago, while at the same time fires raged out of control across Wisconsin, most notably around Green Bay. Although the loss of life in the Peshtigo fire was much the greater, Ingram was more personally touched by the Chicago disaster simply because Ingram & Kennedy did so much business with Chicago firms. Soon letters began to arrive at the Eau Claire office from companies in situations similar to that of the C. L. Rice & Company: "We are entirely destroyed . . . and are obliged to call on our friends and customers to help us in our affliction."[36]

---

[34] Playter to Ingram & Kennedy, October 31, 1871.
[35] Playter to Ingram & Kennedy, October 23, 1871.
[36] C. L. Rice & Company to Ingram & Kennedy, October 12, 1871.

There is something quite indecent about acknowledging advantages which result from another's adversity, however obvious such advantages may be, and so it was in lumber company offices in October of 1871. Aside from the simple observation that lumber prices had suddenly firmed at $16 a thousand, little else was said regarding the Chicago tragedy and its effects on business. As usual, Eb Playter came the closest to offering an honest assessment of the situation, noting that plans to rebuild Chicago would require "quite a pile of Common Lbr."[37] That little of this lumber would be supplied by Chippewa valley producers in no way lessened the advantages. That large amounts of Michigan and Wisconsin east shore lumber would be consumed in Chicago, leaving less to be shipped by rail to the trans-Mississippi markets, was a significant benefit to firms like Ingram & Kennedy.

The winter of 1871–1872 was a good one for lumber sales. Archer left the employ of Ingram, Kennedy & Gill and was replaced at Minneiska by M. B. Richardson, but the change seemed to have little effect on trade at that point. The new manager was soon doing all the business he could handle and had no time for homesickness. "Good sleighing and lumber going fast," he reported in late January. "Forty or fifty teams came in yesterday with wheat and it was a busy time . . . Six or eight teams were on hand this morning at 7½ o'clock waiting for lumber and we were loading last night till 8 or 9 o'clock."[38]

By the end of January, the Eau Claire office had received substantial advances on 1872 lumber, including $5000 from Hearne, Herriman & Company and nearly $10,000 from Dulany & McVeigh, both Hannibal firms.[39] It was apparent that Ingram had succeeded in gaining entry into that market, despite the hardships on the previous season.

Major Day was also enjoying considerable winter trade at Dubuque, although he was becoming increasingly concerned about competition from Minneapolis. Throughout the area, new railroads were offering new opportunities to buyer and seller alike. In the interior of Iowa, with the completion of north-south routes, market areas began to converge and the traditional east-west

[37] Playter to Ingram & Kennedy, October 23, 1871.
[38] Richardson to Ingram & Kennedy, January 23, 1872.
[39] Hearne, Herriman & Company to Ingram & Kennedy, January 5 and January 6, 1872; Dulany & McVeigh to Ingram & Kennedy, January 12, 15, and 27, 1872.

orientation began to weaken. During the winter of 1871–1872, for example, the Burlington & Cedar Rapids Railroad intersected the north-south tracks of the McGregor & Yankton, with the immediate result that Waterloo, Cedar Falls, and other Iowa communities in the neighborhood, formerly the near-private preserve of Dubuque and Burlington lumber dealers, could receive direct rail shipments from Minneapolis. Drummers arrived on the first trains from the north, "offering such inducements as to enable dealers at most points on the Sioux City road to better our prices by purchasing of them."[40]

In order to meet this competition Major Day wasted no time in proposing that they cut their own prices $2 per thousand, but other Dubuque dealers were unwilling to go along with such a drastic reduction. Day was probably correct in maintaining that it would be far easier to discourage Minneapolis lumber in Iowa markets before new arrangements for sales had become established patterns. He was not going to lose a customer for no good reason. Regardless of what the other Dubuque wholesalers might publicly agree to, the Major remained intent upon forcing the new competitors to sell at a loss before they attracted any business away from Ingram, Kennedy & Day. To Ingram he outlined his plans: "To a few of our special customers I will probably give private terms as I can see no wisdom in holding our prices at a point that will force our customers along the line of the Dubuque & Sioux City Road to buy in Minneapolis."[41] Without waiting for a reply from Eau Claire, Day pursued his inclination, selling at figures well below those appearing on the firm's published price list. It was no question of honor; it was a question of business. As Day explained to Ingram, "we are not cutting rates to take any trade away from our neighbors & what we are doing to our customers I am satisfied will be kept confidential."[42]

To offer concessions on prices was not an uncommon practice. Indeed, it is doubtful that many large sales were made at list prices. Then as now, good customers expected special treatment. There was also the practical consideration that by making such private adjustments, dealers could avoid changing their

[40] Day to Ingram & Kennedy, February 4, 1872.
[41] Day to Ingram & Kennedy, February 12, 1872.
[42] Day to Ingram, March 13, 1872.

lists so often. In this instance, the Dubuque firms hoped to reach the first of May without making any published change, at which date it was generally assumed that prices would be reduced considerably.

But Ingram worried less those days about the Major in Dubuque than he did about Kennedy in Eau Claire. This would have surprised most of their acquaintances, since the honest, conscientious, and hard-working Donald Kennedy seemed unlikely to cause much concern of any sort to anyone. Many would also have been surprised to learn that the source of these difficulties was the very success being enjoyed by Ingram & Kennedy. Kennedy had not changed. The business had changed, most importantly in terms of size and scope. While Ingram was steadily moving the firm in the direction of larger production, greater efficiency, and higher profits, his partner worked as doggedly as ever but without the enthusiasm of former days.

At the basis of the trouble was the unmistakable fact that Kennedy was a machinist, not a manager. While his contributions to the efficiency of the sawmilling operation were incalculable, he did not feel competent and was not happy when called upon to assume some nonmechanical responsibility. He had complete trust and confidence in Ingram, and as long as "O.H." was able to handle the details of the business outside of the sawmills, Kennedy was content for him to do so. With the expansion of activities and the increase in volume of business, however, Ingram realized, if Kennedy did not, that some changes had to be made in the organization. Throughout the winter of 1871–1872 he attempted to persuade Kennedy that the time had come to enlarge the partnership, at least to the extent of including bookkeeper-clerk-agent and old friend George Potter, and Ingram's nephew William Tearse, who was then superintendent of the firm's upper or Eddy Mill.

Kennedy simply could not accept the idea of permitting others to share in the decisions of ownership, a reluctance common to many of the old individualists, those self-made owner-operators. The admission that they were no longer able to manage all of the details of a growing business, or that their business might better be served by employing experts at the managerial and staff levels, was not easily made. But times were clearly changing, and the Eau Claire partners were by no means ahead of their time in the consideration of these organizational questions. Indeed, while Ingram was attempting to convince Kennedy of the need for enlarging the partnership, the Wisconsin legislature was discussing a general incorporation bill, which became law in 1872; and Ingram & Kennedy would be the only one of Eau Claire's major lumbering firms which did not incorporate in the decade of the 1870's.

Despite his partner's objections, a number of factors convinced Ingram that the change should not be delayed. In the first place, he recognized the possibility that Potter and Tearse might leave the company unless they were assured some opportunity for advancement. Recalling his own early experiences in Canada, Ingram maintained that the most valuable employees were those who looked forward to a chance to share in the profits of their labors. Also, by rewarding ability and effort the firm would more likely be able to retain the services of those whose contributions were especially valuable and important. Surely in the majority of instances it hardly seemed reasonable to expect any sacrifice or even much personal interest from those having no share in the ownership; and Ingram further contended that at the moment it was more economical to offer some fractional partnership rather than an increase in salary.

But Kennedy remained steadfast in his opposition to any such changes in the organization. In January, 1872, he had taken his son to Boston for a series of operations necessitated by a serious accident. Perhaps that situation was conducive to earnest reflection, for Kennedy used the occasion to answer, in writing, Ingram's proposed inclusion of Potter and Tearse in the partnership.

> With regard to having the boys take an interest in the business, I must say that I don't like the idea at all. I may be wrong in it, but the more I think of it the less I like it, not that I have any fault to find with any of them but I don't like the idea of so many partners; it would be impossible for all to think alike at all times. It is true that having the largest interest we could control them, but men controlled against their will don't feel pleasantly about it as a general thing.

Kennedy did not enjoy disagreements of any sort. It was no easy task for him to tell Ingram, his long-time friend and partner, that if it were possible he would now gladly sell out his interest in the company and look about for some less demanding opportunity, "something that I could make a living at more quietly; even if I did not make as much money." He concluded his Boston letter with the following observation: "I may be mistaken but I don't think that there ever will be much pleasure in lumbering on the Chippewa."[43]

It is not difficult to feel sympathy for Kennedy, who was so very different from his tireless partner and was involved in a situation quite unlike that which he had expected to develop. Kennedy longed for a chance to sit back and enjoy the success already achieved, while Ingram seemed compelled to expand and extend the business, working constantly to improve its position. Although Ingram was not quite as ready as some of his fundamentalist neighbors to equate idleness with ungodliness, rest and relaxation did not come easily to him. Rather, the joys of his life were somehow more the result of making the effort than in the accomplishment. In many respects, Ingram seemed to fit the pattern of the Americans described by de Tocqueville, "so restless in the midst of abundance"—and Donald Kennedy had as much trouble as the Frenchman in understanding such behavior.

But the objections raised by Kennedy could only delay Ingram's plans, and by 1873 both Potter and Tearse were partners in the new organization of Ingram, Kennedy & Company. Their fractional interest amounted to one-eighth each, totalling some $8500 apiece, most of which Ingram loaned them. He made few wiser investments.

Ingram was not always so certain of his own wisdom as he was in the matter of adding Potter and Tearse to the partnership. Although in 1871 he was convinced that an effort must be made to gain entry to the Hannibal market, in the following year he began to have some second thoughts on the subject. He had expected low prices on the lumber while they were establishing contacts and making initial arrangements; but when the low prices persisted into the 1872 season, he wondered whether they should continue to send rafts down to Hannibal, especially since Dubuque seemed ready to consume all of the available Ingram & Kennedy lumber.

It may have been that Ingram was simply unaccustomed to

[43] Kennedy to Ingram, January 26, 1872.

dealing with such noisy and demanding bargain hunters as those in the market at Hannibal. If this was true, however, he learned in a hurry. Even though prices remained low, Ingram & Kennedy lumber was no longer an unknown article. Early in the summer of 1872, the Hannibal customers indicated that they wanted considerable quantities. Dulany & McVeigh, for example, requested a million feet per month, and by the first of July, Dan Dulany expressed regret that they had not ordered an additional 500,000 feet to be delivered each month. All in all, Ingram contracted 10 million feet with Hannibal parties for late summer and early fall rafting, which was nearly half the total output of the Ingram & Kennedy mills in Eau Claire.

As a direct result of this increasing commitment to customers in Hannibal, Dubuque was neglected. There just was not enough lumber to go around, and Major Day unhappily watched the stock in the Dubuque yard disappear while awaiting impatiently the infrequent deliveries from Eau Claire. In mid-July he warned Ingram that unless immediate attention were given to the requirements of Ingram, Kennedy & Day, there was no possibility that they would be able to take advantage of the demands of the fall trade. The Major further noted that they had been burdened by unreasonable handicaps almost since the commencement of the 1872 season, having been forced to ship from Dubuque "largely green lumber to our customers & they are becoming dissatisfied."[44]

If anything, this situation grew more serious as the season wore on. Day seldom addressed Eau Claire without complaining of receiving less lumber than was needed to keep stocks complete and far less than necessary for the purpose of building an inventory which could support active trading after the close of rafting. In September he reported that Ingram, Kennedy & Day had purchased 250,000 feet of lumber from Dubuque competitors in order to fill his own orders. By October he had lost nearly all patience, and when Potter left him only seventeen cribs out of a raft bound for Hannibal, Major Day could not help but wonder if Dubuque still figured in Ingram's plans.

With as much directness as their relationship permitted, Day wrote to Ingram, unburdening himself of an accumulation of

[44] Day to Ingram, July 11, 1872.

grievances. Referring to the seventeen cribs just received, he complained that they had been "similarly disappointed several times this summer. . . ." Day did not quite accuse Ingram of intentionally sabotaging the efforts at Dubuque, allowing that "The difficulties the yard has labored under . . . are possibly more than you have been aware of and probably more than you have designed."[45] Ingram could have no doubts, however, as to the substance of Day's objections.

In the opinion of the Major, he had not received fair consideration in the distribution of Ingram & Kennedy lumber that summer, and it was then too late to make up for the previous neglect. Lumber piled late in the season cost more to handle, and was not in satisfactory condition for the fall, winter, and spring trade, largely because it was not dry enough. Dealers commonly maintained, and Major Day agreed, that lumber piled before the first of October was worth fifty cents to a dollar per thousand feet more than lumber piled after that date.

That, however, was not a basic issue. The Major was most concerned about his role within the organization and the organization's responsibilities to the Dubuque yard. In effect, he demanded that he be allowed to manage the business in Dubuque with the assurance of a fair opportunity for success. He was unwilling to be wasted. He knew that no matter how capable and ambitious and conscientious he might be, a lumber dealer who had no lumber could hardly be expected to show a return. If Ingram had any doubts regarding the seriousness of Day's dissatisfaction, they were dispelled the following winter, when the Major expressed a desire to dissolve the partnership. Only Ingram's acknowledgement of the mistakes made during that season prevented the dissolution.

Mistakes were indeed made. In 1872, Ingram learned the hard way what happened when more lumber was promised than could be provided. In addition to the demands made by Ingram, Kennedy & Day in Iowa, by Ingram, Kennedy & Gill in southeastern Minnesota, and by the parties in Hannibal, regular customers like W. S. Benton of Anamosa, Iowa, and D. T. Robinson of Rock Island had been assured the delivery of ample amounts of Ingram & Kennedy lumber. But the rafts did not arrive, and by the mid-

[45] Day to Ingram, October 7, 1872.

dle of September, with the days growing shorter and cooler, these dealers had begun to get uneasy. With the end of rafting fast approaching, urgent letters reached the Eau Claire office, demanding as Robinson demanded, "now don't fail to send me 200,000 at least every ten days."[46]

During the better seasons, Ingram dealt almost exclusively with the old customers, whose needs for lumber substantially increased at such times. The reason for limiting sales to regular customers was simple enough: it was understood that these dealers would receive preference in the delivery of Ingram & Kennedy lumber in return for the use of any money they might have in surplus.[47] To some extent this system replaced the former practice of advancing money on a specified amount of lumber to be delivered. The new arrangements were considerably less formal, generally implying that if the wholesale dealer had an option or privilege on lumber, the sawmill operator had an option on any surplus funds accumulated by the dealer. But in 1872 Ingram was too often embarrassed in holding up his end of the arrangement.

Because of these recent experiences, Ingram was not overly disappointed when he received notice from Dulany & McVeigh of Hannibal cancelling their plans to purchase for the coming season, although Dan Dulany assured him that "we do like your lumber; and like you."[48] Some of the Hannibal dealers clearly felt that they had more lumber already on hand than the financial situation warranted. Indeed, lumber markets all along the Mississippi appeared to open the 1873 season with small demand and low prices. But a revived Major Day enjoyed a good month of May at Dubuque, while considerable margins were maintained on active trading at the yards of Ingram, Kennedy & Gill.

William Gill did have some problems, however. Of greatest immediate concern, at least to family and friends, was a persistent cough which caused the Wabasha manager increasing discomfort. In April the planing mill was damaged by a small fire which was fortunately brought under control before any serious losses resulted. Almost as soon as the fire had been extinguished, manager Richardson at Minneiska informed Gill that he had received an offer for a better position in St. Louis and was there-

[46] D. T. Robinson to Ingram & Kennedy, September 17, 1872.
[47] W. S. Benton to Ingram & Kennedy, September 24, 1872.
[48] Dulany & McVeigh to Ingram, February 11, 1873.

fore anxious to leave. As his replacement, Richardson recommended that they hire Charles L. Chamberlain, then employed in the wheat warehouse at Minneiska and already well acquainted with the farmers of the area. This suggestion was followed, and Chamberlain assumed his new responsibilities at Minneiska in the middle of May at an annual salary of $700.[49] Gill had long been troubled about the many changes in personnel at Minneiska, but with the appointment of Chamberlain that problem ended.

During the spring of 1873, water ran unusually high on the upper Mississippi and its tributaries, with the result that lumber reached downriver markets with relative ease and in large amounts. The raft pilots, seldom satisfied with any stage of the river, claimed that as much or more judgment was required in running with the banks full as during periods of low water. True or not, on high water the lumber got to market. Ethan Allen, then in charge of running the Ingram & Kennedy rafts below LeClaire, reported that the "Wisconsin river is discharging all she has in the shape of lumber." From a consideration of navigation, the Wisconsin was without question the most difficult of the major streams flowing to the Mississippi from the pineries. When rafting operations on the Chippewa were laborious, they were likely to be impossible on the Wisconsin, and only in part because of the greater distances involved. But in 1873 even the Wisconsin River lumbermen were getting their rafts out in such numbers that pilot Allen could not recall having seen so many ever before: "the river and Islands are litarly filled with them and [there is] no market."[50]

Allen may have exaggerated slightly about the amount of lumber heading toward the markets, and may have been overly pessimistic about the condition of those markets. In any event, Ingram apparently had no difficulty locating customers. By the end of June he had heard from most of the parties in Hannibal, each of whom urgently requested arrangements for more lumber. J. J. Cruikshank inquired whether or not he could be supplied with 4 million feet and Ingram agreed to guarantee half that amount and more if possible at $14 per thousand, one-half cash and the balance in thirty and sixty days. A similar contract was

[49] Richardson to Ingram, May 9, 1873.
[50] Allen to Ingram & Kennedy, June 13, 1873.

arranged with D. Dubach & Company. Even Dulany & McVeigh, regretting their earlier cancellation, sought to obtain delivery of 2 to 3 million feet.[51] But Ingram made his allocations with greater caution than had been the case the previous season, and whenever he approved a contract, he probably felt Major Day looking over his shoulder. Although they were now the second-largest producer in Eau Claire with an annual output of nearly 25 million feet,[52] that amount could be committed in a hurry when supplying points such as Dubuque and Hannibal.

Day and Ingram had resolved their differences, largely to the Dubuque manager's satisfaction. In the wake of his newly won authority, the Major contemplated the construction of a drying kiln at the Ingram, Kennedy & Day yard. This would allow him to ship lumber out of the yard within a week or two after being pulled from the river, thereby making the date of raft delivery of less crucial concern. Also, kiln drying was a much more dependable method than air drying. He estimated the cost for such an improvement to be between $1000 and $2000, a small figure when compared to the advantages which would result, or so he advised Eau Claire. His activities focused on volume, always volume; and, evidencing little concern with prices, Day confidently predicted in the middle of June that they would "make some money during the next year dating from now."[53]

Basic to the Major's plans for profits was the reduction of his own yard expenses per thousand feet, and this in turn depended on an increase in the amount of lumber handled. As he explained to Ingram: "We are now making a fair profit on lbr and have a good trade. I think if we have the stock we can sell 4 million ft this year after this month [June] and that it would be to our advantage to go into winter quarters with a larger stock than last year, say 4 million feet." In at least partial recognition of interests and commitments outside of Dubuque, Day concluded his letter in an understanding and tactful manner. Thus Ingram

---

[51] Ingram to Cruikshank, June 10, 1873; D. Dubach & Company to Ingram & Kennedy, June 11, 1873; D. M. Dulany, Sr., to Ingram & Kennedy, June 24, 1873.
[52] *Chippewa Valley Business Directory for 1873* (Eau Claire: Free Press Print, 1873), 5.
[53] Day to Ingram, June 17, 1873.

read: "I would like to shape matters to this end if you can spare the needed capital."[54]

There were other commitments, and they were not always fulfilled with ease and to the satisfaction of all concerned. Eb Playter often found himself caught in the middle as he attempted to supply a large number of customers without neglecting any one too seriously or for too long a time. There was, however, no possibility that he could keep everyone happy, and in early June it was Major Day's turn to be "fierce." His immediate cause for complaint had been Playter's decision to drop 300,000 feet of a Dubuque raft at Guttenberg, Iowa, for two old customers, B. H. Lueck and Henry Dortland. And so the summer went. Playter's task of keeping customers supplied and content was an impossible one, and he disliked being the object of their displeasure. Even the Fourth of July failed to improve Playter's spirits. He worked all day and celebrated by getting soaked in the rain; and, after putting on some dry clothes, "fell in the River all over. . . ."[55]

One of the major sources of irritation at not receiving rafts at the time expected was connected with pulling lumber from the river. Although this was by no means skilled work, crews varied a good deal in their efficiency, and it was a disagreeable situation for Mississippi buyers to await the arrival of a raft with an idle crew of pullers. Ideally rafts arrived in a manner which would keep the crews steadily at work, so that as soon as they had completed pulling, washing, sorting, and piling the last crib of one raft, another would be in sight. But with the large demand for lumber and the dependency of so many dealers on Ingram & Kennedy deliveries, some delays and corresponding disappointments could not be avoided.

In July, D. T. Robinson made the expected objections when Playter announced that the raft then passing Rock Island had all been marked for Hannibal. Complaining that he had ordered two strings some three weeks earlier and should long since have received his lumber, Robinson proceeded to recall what seemed to be a series of misfortunes in his dealings with Ingram, Kennedy & Company. He reminded Ingram that "last fall [I] was annoyed almost beyond endurance [and] now it is repeated." As

[54] Day to Ingram, June 19, 1873.
[55] Playter to Ingram & Kennedy, July 2 and July 4, 1873.

for future deliveries, the Rock Island dealer requested that if he could not possibly be accommodated on any convenient schedule, he preferred that his lumber be allowed to come soon and all at once if necessary, adding, "I had rather be troubled that way than have men standing around waiting. . . ."[56]

This problem was caused by a combination of factors, some of which were beyond control. Even assuming that the sawmills of Ingram, Kennedy & Company could produce lumber at a rate equal to demand, Eau Claire was many miles and many obstacles away from the levees of Dubuque, Rock Island, and Hannibal. Yet these difficulties, however unavoidable they might have been in their origin, were clearly complicated by the bookkeeping and clerical procedures of the Eau Claire office. No one realized better than Orrin Ingram—unless it happened to be his ex-partner Dole—that office management was not one of his strengths. The responsibility was his largely by default, and indeed he had long wished for an assistant who could assume charge of the books and handle the daily correspondence. But these were the kind of additional costs, since they were not directly involved in the production of lumber, which tended to be postponed as long as possible. Nevertheless, it was clear that the increasing amount of paperwork which paralleled the increasing volume of business was rapidly becoming more than Ingram could manage alone. Furthermore, a system was required, the adoption of some standard procedures. A good memory and figures scratched on the back of an envelope no longer sufficed. Customers were becoming aware of the clerical deficiencies of the Ingram, Kennedy & Company office, and occasionally they would offer suggestions as to possible improvements.[57]

[56] Robinson to Ingram & Kennedy, July 21, 1873.

[57] Perhaps the most important of these suggestions came from John J. Cruikshank, Jr., of Hannibal, who complained on July 17, 1873, that he was forever having to remind Ingram of what they had previously agreed upon. In reviewing the details of the contract they had negotiated earlier in the summer for the delivery of 2 million feet of lumber, Cruikshank finally sent a copy of Ingram's own letter of July 10 back to Eau Claire along with a note addressed to "Friend Ingram," which read in part: "If I thought you would not think I was a presumptuous youth I would advise you to copy all your letters. I can't imagine how you can do so much business . . . and have gotten along so long without a copying Press. I think if you will try it for a while you will wonder how you have got along with out one and I know if you try it you will always Copy."

But improvement in office procedures could not eliminate all of the problems associated with lumber distribution. Although Major Day's situation at Dubuque was considerably better than the previous season, he still found occasional opportunities to complain about what seemed to him an insufficient supply of Ingram, Kennedy & Company lumber. In late August, for example, when George Potter attempted to leave him only thirty-one cribs out of a raft headed for Hannibal, the Major insisted that he be given more. Day reported the disagreement to Ingram, observing that had he accepted Potter's original assignment of thirty-one cribs, he would have been forced to discharge half of his piling crew. The delivery of the fifty-six cribs finally agreed upon, however, allowed him to let just three or four men go and to keep the remaining crew members hard at work. "I am very anxious to keep together my present efficient crew," Day explained, "until we have piled most of what we intend piling this season."[58]

The costs associated with pulling, washing, sorting, and piling were not constant, and dealers such as Day naturally tried to avoid any extra and unnecessary expenses for such work. Per-unit costs in the handling of lumber could be reduced considerably once an efficient, hard-working crew had been organized and trained. To be compelled to pay them for no work or to dismiss them for lack of work, only to hire a new crew when the lumber arrived, was an aggravating alternative. These subjects were of particular concern to Major Day, so conscious of the importance of doing business on a large volume. Costs had to be kept at the very minimum.

Lumber prices remained firm at $14 per thousand feet at Dubuque through the month of August, but rafts arrived at the Ingram, Kennedy & Day levees at an accelerated rate. Major Day wanted to take full advantage of the good weather, and wanted to get as much lumber in condition for fall and winter shipment as could possibly be managed. In past seasons there had been too much green or "heavy" lumber leaving the yard to suit the Major and to please his customers. Even should Ingram determine that it was not possible to provide the Dubuque yard with the quan-

---

[58] Day to Ingram, August 21, 1873.

An early view (ca. 1870) of a portion of the Ingram & Kennedy property, with West Eau Claire providing the bleak backdrop. The Ida Campbell *was used for many years in rafting operations along the Chippewa.*
(State Historical Society of Wisconsin)

Orrin H. Ingram, from a daguerreotype dated 1857, the year he arrived in Eau Claire to begin his independent lumbering operations.
(Courtesy Edmund Hayes)

Cornelia Ingram, wife of Orrin, whose life centered principally around church-related activities.
(Bartlett Collection, Eau Claire Public Library)

Aerial view of Half Moon Lake. Although this photo was made many years after lumbering ceased in the Chippewa Valley, the geography remains as before. The Chippewa can be seen winding away into the distance.
(State Historical Society of Wisconsin)

A portion of the Dells improvement on the Chippewa, just above Eau Claire. This photo was probably made in about 1890, during the spring when the logs had begun to arrive from upstream.
(Courtesy Edmund Hayes)

*Portion of the Empire Lumber Company facilities along the Chippewa,
ca. 1890.* (Bartlett Collection, Eau Claire Public Library)

*A good-sized load of pine logs heading for
the landing at one of the Empire camps.
Probably these loads tended to be a bit
larger when photographers were in the
woods.* (Courtesy Edmund Hayes)

Panorama of the Empire facilities. At left center is the bull slide on which logs moved from the river up into the sawmill. In the center is the shed where rough lumber was made into cribs. Angling from the shed into the river are the ways, down which the cribs were launched. And in the fore-ground is a lumber raft. (Courtesy Edmund Hayes)

Orrin H. Ingram and Frederick Weyerhaeuser (third and fourth from left, front row) during a tour of pineries in the South during the 1890's.
(Courtesy Edmund Hayes)

View of the mouth of Beef Slough, about 2½ miles above Alma, Wisconsin, on the east bank of the Mississippi. For a time this was the busiest log harbor in the United States, employing between 500 and 600 men during the log-driving season. (State Historical Society of Wisconsin)

Beef Slough, catch basin for logs of Weyerhaeuser's "Chippewa Pool." In these works, logs from the Chippewa watershed were impounded, scaled, and then combined into huge rafts for the continuation of their journey down the Mississippi. (State Historical Society of Wisconsin)

Lumber raft with an unlikely crew—though in fact it was not uncommon for family and friends to enjoy a ride downriver when the weather and stage of water were good. The steamboat in the background is pushing the raft. (Courtesy Edmund Hayes)

An unusual view from between the stacks of the pushing steamboat, towards the bow boat which is lashed parallel to the front of the raft. By going forward or backward, the bow boat could impel the raft to left or right.
(State Historical Society of Wisconsin)

Reads Landing, on the Minnesota side of the Mississippi opposite the mouth of the Chippewa, was to lumber what Beef Slough was to logs. Here the Chippewa River rafts were made up into much larger Mississippi rafts for movement to downriver markets. (State Historical Society of Wisconsin)

Contemporary view, ca. 1890, of two lumber rafts on their way down the Mississippi. Gerhard Gesell made this photograph from the bluffs above Alma, Wisconsin. (State Historical Society of Wisconsin)

The Ingram home in Eau Claire, adjacent to the church which played a prominent role in the Ingrams' lives.
(State Historical Society of Wisconsin)

Grandfather Ingram (left) and friends on their way to the summer cottage at Long Lake.
(Courtesy Edmund Hayes)

Orrin H. Ingram in his later years.
(Courtesy Edmund Hayes)

tity desired, Day wanted it clearly understood that he would "much prefer the supply for the yard . . . be cut off at a later period when the days are shorter & men work to poorer advantage on that account and lumber has not time to season before freezing." Day's preferences in this matter were hardly unique. Nevertheless, Ingram made every effort to accommodate his Iowa partner, and by September 1, 1873, nearly 3½ million feet of lumber were piled in the Dubuque yard of Ingram, Kennedy & Day, an increase of 600,000 feet over the previous year. Business was good and showed signs of further improvement.[59]

Even friend Benton, a near-chronic sufferer from the "blues," thought that the business outlook in the Anamosa region of Iowa seemed promising. He was certain that he could make some money "if the derned fools of Farmers will only let the Granger business go and atend to their business, and sell their grain now . . . but they are making the bigest fools of themselves of any set of men I ever saw."[60] Concern with matters such as the Grangers was soon overshadowed by a much larger threat to the national economy. Shocking news of the failure of Jay Cooke & Company, followed by similar reports involving other famous banking houses and then by the closing of the New York Stock Exchange, suddenly provided plenty of fuel for fear. But as Ingram happily discovered, reactions to the new panic were by no means predictable in the lumber markets of the west, and by no means uniform.

In the week following receipt of the bad news from Wall Street, Ingram heard from all of his Hannibal customers, each requesting that no more lumber be rafted to their market. When Ingram suggested that they should settle their accounts on lumber already delivered, he received letters describing the hard times but containing no money. D. Dubach & Company offered a simple explanation: "Have not a dollar." John B. Price answered Ingram's telegram which had urged the remission of some currency by observing, "That little thing has played out in this place. . . ."[61]

---

[59] Day to Ingram, August 27 and September 3, 1873.

[60] Benton to Ingram & Kennedy, September 13, 1873.

[61] D. Dubach & Company to Ingram & Kennedy, September 27, 1873; John B. Price to Ingram & Kennedy, September 27, 1873.

The Panic of 1873 lost no time in arriving in the Mississippi valley. On September 26 the Merchants National Bank of Dubuque suspended operations. Subsequent investigation seemed to indicate that dishonest management contributed as much to the closing as anything else, but at the moment, the failure was another indication to Major Day that it was time to retrench and prepare for the worst. Like the customers in Hannibal, he too wanted no more lumber, advising Ingram that "Until this panic is over & business has resumed its usual course I am inclined to think it would be better for you to realize out of your future rafts by sale in the water."[62]

What had been feared, however, did not develop, at least not everywhere and certainly not immediately. During the month of October, Major Day reported that the Dubuque yard accumulated $21,000. In the middle of November, Gill acknowledged that collections had been running somewhat behind at Wabasha and Minneiska, but he blamed the condition of the roads rather than the financial situation.[63]

Meanwhile, Ingram proceeded with the arrangements for winter logging operations as if times were normal, although there was not quite the complete freedom in such matters as many assumed. Indeed, some of his wholesale customers were impressed by what seemed to them an unusual display of confidence. Dan Dulany, for example, observed from Hannibal that Ingram was one who never looked "at the dark side of the picture. . . ."[64] Other lumbermen, however, were less charitable in their assessments of the preparations for another big season of logging. Hearne, Herriman & Company, also of Hannibal, thought that if the sawmill operators had any sense at all, they would reduce the supply of lumber rather than increase it.[65] Quite naturally, curtailment became a common cry during bad times; but few things received more attention with less result.

The Panic of 1873 seemed to effect the lumber markets below Dubuque more seriously than those above. In the middle of December, for example, the Winona, Minnesota firm of Horton

[62] Day to Ingram, September 25, 1873.

[63] Day to Ingram & Kennedy, October 28, 1873; Gill to Ingram & Kennedy, November 14, 1873.

[64] Dulany & McVeigh to Ingram, November 25, 1873.

[65] Hearne, Herriman & Company to Ingram, December 19, 1873.

& Hamilton expressed complete confidence that their trade for the winter would be good. Gill was equally optimistic and in early February, 1874, with temperatures cold and sleighing ideal, the manager of the Wabasha yard reported to the Eau Claire office, "Trade was never better," in proof of which Gill enclosed a draft for a thousand dollars.[66] Major Day was also keeping busy at Dubuque. Completion of the stock inventory and closing of the books for 1873 gave evidence of considerable progress over the past season. Ingram and Kennedy's investment in the Dubuque operation had been reduced below $100,000, and the total investment of Ingram, Kennedy & Day in the various retail yards across Iowa had been reduced by nearly one-third to less than $25,000. They also owed $7000 less, were owed $5000 more, and had $13,950.17 more value of lumber in the yard than on January 1, 1873. Day readily acknowledged: "I feel more encouraged for the immediate future than I did last season."[67] Although the profits for 1873 had been figured at slightly in excess of $20,000, this was no fair indication of the success recently enjoyed. More often than not, lumber was inventoried at rates well below its actual value. Funds were also commonly assigned to so-called surplus accounts, to be used in financing improvements; and significant amounts had to be expended, of course, in reducing principals and in meeting interest payments. As a result, what was designated "profits" was often but a fraction of the actual margin of price over cost. From any perspective, 1873 had been a very good year for Ingram, Kennedy & Day.

Business at Dubuque remained active as the winter months passed. By late February, Major Day inquired of Ingram if he might take out a million feet from the rafts that had been run into the slough there for the winter. This temporary disposition had followed the urgent requests from Hannibal the previous September that no more lumber be run through to that point for the balance of the 1873 season. Since the continued large volume of trade at the Dubuque yard had broken up the assortment to a considerable extent, the Major thought he might replenish his stock from the rafts tied up there. He expressed the hope, how-

---

[66] Gill to Ingram & Kennedy, February 5, 1874.
[67] Day to Ingram & Kennedy, December 30, 1873; balance sheets for Ingram, Kennedy & Day, 1873.

er, that Ingram, Kennedy & Day would be billed no more than 2.50 for any lumber he was allowed to appropriate from the Hannibal rafts.[68]

Although business remained better than many had expected, the Eau Claire office of Ingram, Kennedy & Company did receive fewer advances during the winter on 1874 lumber than had become customary. There was almost no expression of interest from the parties in Hannibal. Also, on the advances that were received, the prices arranged were definitely lower than those that had ruled for the past few seasons. For example, Ingram agreed to deliver some lumber to B. H. Lueck of Guttenberg, Iowa, for $12 per thousand. But Guttenberg was one of the first stops below Reads Landing, and to one who had so recently sold large amounts at Hannibal for $14, the price to Lueck could hardly seem disastrously low. Anyway "the Dutchman" was an old friend and steady customer, even if his letters had to be read aloud to be understood. He had written to Eau Claire in mid-February that he was busily selling "mohr Lumber as all Yards hier in Guttenberg," adding, "mein Fensing ist all Sold." In early March he asked Ingram to "plies let mi noh wat teim yu kom doun mit Lumber."[69] Major Day was not the only dealer anxious for rafting to begin.

The arrangements made with Lueck, however, proved to be an exception. Ingram saw no need of agreeing to twelve-dollar prices for his lumber. Indeed, he was sufficiently confident of business prospects that he finally made that long-delayed appointment to the office staff, hiring twenty-eight-year-old Clarence Abner Chamberlin as chief clerk for the company. Chamberlin's experience had largely been acquired clerking on Ohio and Mississippi river steamboats, but he had also spent a little time working in an Eau Claire furniture store. Ingram had seen and heard enough to know that the young man was honest and diligent and had acquired some of the basics of bookkeeping. There is no question that the future success of Ingram was, to a very considerable degree, the result of Chamberlin's efficiency. Ingram and Chamberlin would be associated together in business until the latter's death in 1911; and if Ingram ever had

[68] Day to Ingram, February 23, 1874.
[69] Lueck to Ingram & Kennedy, February 16 and March 2, 1874.

a close friend, it was Clare Chamberlin. In the spring of 1874, however, the new clerk found that he had inherited a very confused desk in what was rapidly becoming a large and increasingly complex organization.

Ingram's confidence that the demand for lumber would persist in spite of the depression proved correct. As the rafting season approached, even F. P. Hearne of Hannibal admitted that "Trade is getting pretty good," although he quickly added that there had been no advance in prices.[70] But conditions must have improved or else never were as bad as earlier portrayed by the Hannibal crowd and, by late May, Dan Dulany decided that it was time for him to go to Eau Claire to discuss trade possibilities with Ingram in person.[71] The old Kentucky colonel was not one to make such a long trip unless he was extremely anxious to arrange for some lumber.

If the prices on lumber held at slightly lower figures during the 1874 season, there was no appreciable difference in the amount of trade. At Dubuque, for example, Major Day faced the same problems of insufficient supply. In fact, the lower prices served only to increase his commitment to volume sales, as again he explained: "At the present low margins on lumber our only hope for profits consists in being able to handle a large quantity and I am especially desirous that we shall have an opportunity of trying to succeed in this manner." While he recognized that the Eau Claire office was under pressure to keep many different dealers supplied, the Major warned that unless his assortment in the Dubuque yard was adequately maintained so as to satisfy Ingram, Kennedy & Day customers, "there will probably soon be all the lumber we can handle."[72]

Major Day's problems, however, did not simply involve the amount of lumber marked for Dubuque delivery. Actually, keeping the inventory current in the yard had become the more fundamental difficulty, as well as keeping the mills in Eau Claire informed of his specific requirements. But distance and the uncertainties of communication combined to make the best efforts along these lines inadequate. As a result, all too often Day was

---

[70] Hearne, Herriman & Company to Ingram & Kennedy, March 26, 1874.
[71] Daniel Dulany, Sr., to Ingram & Kennedy, May 29, 1874.
[72] Day to Ingram & Kennedy, May 29, 1874.

forced to purchase a particular kind or size of board from his Dubuque competitors if he was to fill his customer's order exactly. Although both Day and Ingram knew what the solution to the problem was, both avoided any direct reference to the construction of a sawmill, for the time being anyway. In the middle of June, the Major did recommend that they purchase "a few hundred thousand feet of the cheap Black River logs," and have them sawed to order by mills in Dubuque. Within a month Day had his Black River logs and was confident that they would serve the purpose.[73]

During the first six months of 1874, the Dubuque yard sold 5½ million feet of lumber and had about 3 million feet remaining in stock. Looking ahead to the balance of the season, the Major thought it likely that they would sell another 5 million feet and that they should "go into winter quarters" with no less than 6 million feet piled. In order to arrive at such a disposition, from the first of July until the end of rafting they would have to yard some 8 million feet. If they figured on 110 working days, this would mean that the Ingram, Kennedy & Day pullers must handle an average of 70,000 feet per day. The Major was not overly confident that Ingram would consent to such a rate of supply, admitting that his request might be termed extravagant. But they were burdened with a "voracious yard," as Day described the Dubuque operation; and he reiterated his belief that "our only opportunity to reap a liberal success on the low margin of profits we are reduced to is to handle larger quantities than heretofore & to this end I am working and will only be deterred by lack of stock."[74]

A week or so later, the Major increased his estimated requirements to 10 million feet as his trade continued to move along at a brisk rate.[75] By the end of August the Dubuque yard held more than 7 million feet and the piling crews were still going strong. The demand for lumber had diminished to some extent, but Day remained confident that he could handle 3 million more feet to good advantage. He explained the slowdown in trade by the low price of wheat. As usual during such periods, farmers

[73] Day to Ingram, July 11, 1874.
[74] Day to Ingram & Kennedy, June 21, 1874.
[75] Day to Ingram, July 11, 1874.

seemed intent on delaying sale of their grain as long as possible, hoping for an improvement in the price they received. In the meantime they paid only their most pressing bills, "largely giving the lumber merchant the slip."[76]

Demand for lumber fell off generally in the western markets in September, and consequently Ingram was more than ready to provide Day with all the rafts he could handle. The Major had no objection to being a safety valve for lumber that had suddenly become surplus, and he indicated his readiness to accept "an unusual quantity" provided they could agree on a fair billing price and find a sufficient area for the piling of the additional lumber. But he also reported that C. Lamb & Sons and W. J. Young & Company, large producers in Clinton, were then selling common lumber at $2 below list prices, or at about $11 per thousand. Lumber was still costing too many bushels of wheat, however, and farmers continued to hold out for higher prices on their grain. Day admitted that he had been caught with a good many outstanding accounts. Yet, during such periods, there were definite limits on the pressure that could be applied to make collections. Not uncommonly a bill might be settled, but at the cost of losing a customer.[77]

In Anamosa, Benton tended to be less charitable than Major Day concerning their farmer customers, though one suspects that Benton's letters to Eau Claire contained considerably more venom than his conversations at home in the yard. Asking Ingram's patience, Benton apologized for being so tardy in paying on his Ingram, Kennedy & Company account and he proceeded to describe the difficulties he had encountered in trying to collect on bills due him. "These – – – Grangers have got to be the meanest men on the face of [the] earth," the Anamosa dealer contended. Citing their determination not to market wheat at present prices, Benton characterized the Iowa farmer as being "as independent as a hog," although he still held a bit of hope that "they will soon see the folly of it and become *better men*."[78]

Nearly a month and a half passed before Benton wrote again, but the passage of time benefited neither his business nor his outlook. Although he noted that wheat prices had fallen to sixty-

[76] Day to Ingram, August 28, 1874.
[77] Day to Ingram, September 20, 1874.

five cents a bushel in Anamosa, he maintained that most farmers were just using that as an excuse to avoid paying their debts. The real trouble was that "we have spoilt them . . . and now we are getting our pay for it." Rather than paying their bills like good people do, the Grangers "just tell us we have got to wait until they get ready to sell." Benton allowed that he was tired of waiting. He had commenced several law suits and was fully prepared to commence more if necessary, explaining that he had decided "it is our duty to learn them a lesson on Business and [I] shall do my part toward it."[79]

Benton's problems with the Grangers were apparently largely local if not strictly personal. Good times had returned to Dubuque, where October sales for Ingram, Kennedy & Day amounted to nearly $30,000, or more than $7000 in excess of any previous month.[80] Ingram was unquestionably buoyed by the Major's continued success, and in a rare moment of generosity he offered a rebate on the late season deliveries to Dubuque to make matters "come out all right" with the yard. Although Day had never doubted that they would realize a fair profit, if only because of his increased volume, since Ingram had raised the subject, the Major suggested in reply that the Ingram, Kennedy & Company lumber "be fairly billed at $11.50."[81]

Ingram had also been encouraged by the likelihood of establishing an increased trade with Horton & Hamilton of Winona. He and Charles Horton had gotten well acquainted in recent months, and all indications were that they would become even more involved in the near future. Approximately the same size as the Dubuque operation of Ingram, Kennedy & Day, Horton & Hamilton had yarded some 11 million feet during the 1874 season, only a fraction of which had been provided by the In-

---

[78] Benton to Ingram, Kennedy & Company, October 1, 1874.
[79] Benton to Ingram, November 9, 1874.
[80] Day to Ingram, November 6, 1874.
[81] Day to Ingram, October 22, 1874. By "fairly billed" the Major simply meant that the price established should be competitive. Although Day and Ingram were partners at Dubuque, the Eau Claire operation was another matter, and, correctly or otherwise, Day occasionally worried that Ingram preferred to show the balance at the sawmill. Anyway, pricing lumber was a point of normal contention because, regardless of the ownership arrangements, such prices would determine for the record which function made the profit (or loss), a determination which was important for all manner of considerations.

gram, Kennedy & Company mills. Looking towards the commencement of rafting in 1875, however, Horton told Ingram that his Winona firm would be requiring "at least 4 to 5 million of your lumber next season."[82]

Considering the new market possibility in Winona, together with the certainty that Hannibal customers would soon be in need of substantial amounts of lumber, during the winter of 1874–1875 Ingram agreed that the time had come to provide alternate means of keeping Day supplied at Dubuque. It was already too late to undertake construction of a sawmill for the 1875 season, but clearly something had to be done. Despite the quantities of lumber which had been piled in the Ingram, Kennedy & Day yard during the season just past, the Major repeated his request of a year earlier; namely, that he be allowed to fill out his broken assortment by pulling cribs from the Ingram, Kennedy & Company rafts which had again used the Dubuque steamboat slough as a winter harbor. In March, 1875, Day again reported sales amounting to nearly $30,000.[83]

As a partial solution to his problem of lumber supply, the Major proposed that they purchase a half-million feet more of the No. 2 Black River logs which he estimated would cost about $7 per thousand. He had contacted his neighbor Charlie Clark, who indicated a willingness to saw the logs according to Day's directions for $2.50 a thousand. Ingram preferred, however, that the Major arrange for the leasing of a sawmill, in part to gain experience for the season when he would have to assume charge of a mill, a day no longer distant. Actually, both suggestions were followed: Clark sawed Black River No. 2 logs for Ingram, Kennedy & Day; and in early June, the Major succeeded in leasing the Iowa Lumber Company sawmill which by August was producing a daily average of 40,000 feet of lumber.[84]

Still these arrangements did not begin to satisfy the total requirements of the Dubuque yard. Day still depended on the Eau Claire mills for most of his lumber. Accordingly, when Ingram observed that unfavorable log driving conditions on the Chippewa might prevent any large deliveries to Dubuque early

[82] Horton & Hamilton to Ingram, November 6, 1874.
[83] Day to Ingram, April 7, 1875.
[84] Day to Ingram, June 4 and August 9, 1875.

in the rafting season, Day confessed "to a sense of disappointment." But after a brief pause for reflection he felt considerably better, explaining to Ingram, "Were you not quite in the habit of performing in excess of your promising I would be quite discouraged."[85]

In line with the business precepts of Major Day, if margins remained it was not possible to sell too much lumber. Had it been possible, he might have managed to do so in the spring of 1875. The volume of trade out of the Dubuque yard for April exceeded 1½ million feet and amounted to more than $31,000, "the heaviest month we have ever had." By the first of May, only 2½ million feet remained in stock, causing Day to warn: "The demands of our trade were never so imperative or the danger of our making a complete surrender to our competitors so imminent as this spring."[86] But Ingram could better recall seasons when lumber went begging, and when he considered the difficulties that might have been, he tended to worry less about the problem of Day's assortment.

Furthermore, Dubuque was not the only point demanding attention. George Potter returned to the river as an agent with the start of the 1875 season, and in one of his first letters back to the Eau Claire office he reported: "Everybody is hungry for lumber." Realizing the limitations of production and delivery, especially early in the season, Potter suggested that Ingram try to purchase some lumber from other sawmills, at least until their own mills were in full operation. There were not many opportunities for profitable arrangements of this sort. Nevertheless Ingram succeeded in contracting for about a million feet with the firm of Olds & Lord, located at Afton, Minnesota, and the price agreed upon was $10 per thousand in the raft at Reads Landing.[87]

Somewhat surprisingly, the parties in Hannibal remained relatively quiet throughout most of the summer of 1875, although they may have realized that Ingram was not anxious to compete in that market with lumber such as that being rafted there by

---

[85] Day to Ingram, April 25, 1875.
[86] Day to Ingram, May 20, 1875.
[87] Potter to Ingram & Kennedy, May 2, 1875; Olds & Lord to Ingram, Kennedy & Company, June 8, 1875.

Daniel Shaw & Company at the low price of $13 per thousand. It was not a question of Ingram giving up on Hannibal, for he was anxious to resume trading there. But at the moment, there seemed insufficient reason to raft lumber much below Rock Island when dealers closer to home were so in need. For more than two seasons now—since September, 1873—the Hannibal customers of Ingram, Kennedy & Company seemed unable or at least unwilling to pay competitive prices for their lumber. For the most part, Ingram was content to wait, although he did offer an occasional reminder, as, for example, when he sold some lumber in the lower market through John C. Daniels at the very end of the 1875 rafting season. In this instance Daniels had reported back to Eau Claire that it was "by far the Best that has been Pulled here this year & I should like to Handle such Lumber as that & I could get more than $13 pr M for that kind."[88] Even if commission dealers like Daniels could be believed, there had been no need for Ingram to press matters in that market with demand so large above. Indeed, combining the requirements of Ingram, Kennedy & Day and Ingram, Kennedy & Gill there had not been much lumber left to worry about.

At Wabasha, however, there was concern about another matter. William Gill could no longer ignore his deteriorating health, and he departed for California in the hope of being cured of what he called "bronchial catarrh." (More probably Gill left Wabasha with an advanced case of tuberculosis, for he died in distant San Diego the following March.) The resulting reorganization of Ingram, Kennedy & Gill was accomplished by shifting Charles Chamberlain from Minneiska to Wabasha, with William M. Perry assuming the management of the smaller yard. But Gill was not forgotten. Ingram refused to change the former division of ownership, thus assuring Gill's widow a substantial income. He also attempted to be of some assistance in the raising of the Gill children, although his success in this endeavor was hardly noteworthy.

In the meantime, at Dubuque there was activity of a different sort during the winter of 1875–1876. Any thoughts of further postponing the construction of a sawmill for Ingram, Kennedy & Day doubtless went up in the smoke of a November fire, which

[88] John C. Daniels to Ingram, Kennedy & Company, November 8, 1875.

destroyed the Iowa Lumber Company mill which the Major had leased in the season just past. Also consumed by the flames were more than 1½ million feet of Ingram, Kennedy & Day lumber, worth an estimated $30,000. Most of the loss was covered by insurance, but in some respects it was difficult to measure the cost of the fire, for, as Day explained: "The derangement of our business for want of this stock, and the prospective profits arising from the entire venture will swell this loss considerably."[89] But the setback was actually minor and soon forgotten by the Major as he followed a busy Kennedy around, offering suggestions about this and that. Slowly, under Kennedy's supervision, a new and modern sawmill began to take shape. Major Day watched the construction proceed with great satisfaction. The commitment to Dubuque was now obvious, and opportunities for the future had correspondingly increased. Always a dandy, he marched through the streets armed with quicker smiles for the gentlemen and deeper bows for the ladies. Fortune was clearly smiling on the Major.

Will Tearse arrived in Dubuque in the spring to assume the responsibilities of sawmill superintendent. The whistle blew for the first time on the morning of June 1, and the Major reported to Ingram a few days later.[90]

> I think we have reason to be gratified at the prospect of meeting with success in building what we undertook to build, viz., a first class mill of the size and description designed. The machinery is certainly working very admirably. The draft to the furnace could not be improved and the boilers make steam with an ease. . . .
> Our location for the mill I am satisfied is the best in the city and I think we will experience no difficulty in handling logs in high water or low. . . .

This initial assessment was a bit overly optimistic. The Major and Tearse would experience many problems and some serious accidents in the course of the next few weeks. But one by one the trouble spots were eliminated, or at least smoothed over, and the efficiency of the mill steadily improved as men and machines warmed to the task. By late July, the daily cut averaged

[89] Day to Ingram, November 19, 1875.
[90] Day to Ingram, June 4, 1876.

60,000 feet. A pleased Major Day thought his mill could "safely be considered in successful operation."[91]

Similar success was not apparent in the marketing efforts. Prices ranged generally lower and demand weaker than had been the case for many years, a situation that would continue for the remainder of the 1870's. In early May of 1876, the usually cheerful Day warned that "with Presidential Election, Centennial, and the hackneyed cry of hard times to contend with" they should all prepare themselves for a disappointing year.[92] In stating his reasons for pessimism, the Major neglected to include reference to the grasshoppers, which, when they finally left northwestern Iowa, took with them about one quarter of the corn crop, or so it was claimed. Lumber prices fell to $11 per thousand, and sometimes less, in the markets of Dubuque and points above, while the Hannibal dealers were not interested in paying much more than $12. But the lumber was sold for whatever it would bring, and in the forests and sawmills lumbermen directed an increasing amount of attention towards reducing costs, thereby maintaining their margins.

Reducing costs proved an especially difficult assignment during 1877 and 1878, however. Loggers operated under great handicaps because of an almost complete absence of snow in the Chippewa pineries. The observation of a Chippewa Falls reporter for the *Northwestern Lumberman* in January, 1878, was repeated many times over: "The weather is delightfully clear and moderately cool. No snow yet. The loggers are sick."[93] Some dealers looked forward to the benefits they assumed would follow on the heels of the curtailment imposed on producers by the unfavorable weather. But in this they were to be disappointed, and if stocks were substantially reduced, the difference was slow to affect price levels to any significant extent.

In spite of all of the difficulties, Ingram, Kennedy & Day sales amounted to well over $180,000 for 1877, and the Major remained confident that he would be able to find buyers for all of the lumber the new mill could produce. In early spring of 1878, he predicted that they would be able to command at least

---

[91] Day to Ingram, July 28, 1876.
[92] Day to Ingram, May 2, 1876.
[93] *Northwestern Lumberman,* January 5, 1878.

fourteen-dollar prices on common lumber from the Dubuque yard. In the meantime, Chamberlain had advanced prices at Wabasha by $2, bringing his common boards back to the fifteen-dollar-per-thousand level.

But advancing prices was not the only activity in Wabasha. The Ingram, Kennedy & Gill organization was in the midst of expanding its operation, following the new Valley Railroad west and opening retail yards in the small rural towns of Millville, Mazeppa, Zumbrota, and Zumbro Falls. Chamberlain thought that their little corner of southeastern Minnesota might consume 8 to 10 million feet of lumber each year—a prediction which proved to be an exaggeration of some 50 per cent. Still, there were good reasons for optimism. Charlie Horton in nearby Winona noted that wheat prices along the road west had returned to levels exceeding a dollar per bushel, and he informed Ingram that Horton & Hamilton needed lumber "very much."[94] Clearly, conditions in the upper Mississippi valley that spring of 1878 were on the mend, at least as concerning the marketing of lumber.

Apparently the situation in Hannibal was slower to improve. In April, Dulany & McVeigh and D. Dubach & Company had each offered to purchase a million feet of Ingram, Kennedy & Company lumber, to be delivered in July at a price of $10 per thousand for the dimension and $12.50 for the boards and strips. Ingram declined, however, replying that the prices offered were simply too low. Upon receiving the rejection from Eau Claire, Colonel Dulany countered by betting Ingram "an *Oyster Stew* that you will sell Lumber in July at the prices we offered you. . . .[95]

Despite this failure to agree on terms, relations between Ingram and the Dulanys were getting closer, and it seemed merely a matter of time before large trading was resumed between Ingram, Kennedy & Company and wholesale dealers in Hannibal. Dealers like Dulany & McVeigh wanted to buy Chippewa lumber and Ingram wanted to supply them, but he would no longer do so at figures that were less than competitive with other markets. He could hardly accept an offer of $10 and $12.50 from Hannibal parties at the same time he was receiving bids of $11

---

[94] Horton & Hamilton to Ingram, Kennedy & Company, May 1, 1878.
[95] Dulany & McVeigh to Ingram, Kennedy & Company, April 19, 1878.

and $13 from firms such as Burch, Babcock & Company of Dubuque, regardless of what prices might be in July.[96] Major Day made specific objection to the contract arranged with Burch, Babcock & Company, one of Ingram, Kennedy & Day's major competitors, but in the Eau Claire office customers were customers.

Another problem which complicated matters in the spring and early summer of 1878 was the uncertainty of the log supply and the difficulty of getting lumber rafts out of the Chippewa. It was an unusual period, with too little rain in the pineries and too much on the prairies, although log shortages, like crop damage, were seldom as severe as advertised. Nevertheless, low water on the Chippewa caused the lumbermen considerable worry and no small expense. Until the logs reached the Eau Claire booms, caution had to be exercised as to the amount of lumber contracted for with downriver customers.

Following the earlier failure to reach agreement with Ingram, Colonel Dulany journeyed up to Eau Claire, and upon his return to Hannibal he and Dubach made another offer, this time for 5 million or more feet at a price to be determined at the time of delivery. When he received no reply from Eau Claire the Colonel wrote again, indicating their continued interest in receiving "several million ft. of your lumber during the season," and enclosing a draft for $5000 by way of an advance.[97] An answer was shortly received, but not what Dulany and Dubach had expected. Ingram again refused to promise delivery in any amount, not because of price considerations but because of a shortage of sawlogs in Half Moon Lake. He had learned by trial and error that there was no advantage in contracting for more lumber than could be provided; not only did you make no money, you ran the danger of losing customers.

The difficulty experienced at Eau Claire in the summer of 1878 in securing a sufficient supply of logs was by no means unique. Indeed, the consequences of the extended deficiency of snow and rain were apparent throughout the pine regions of the upper Mississippi valley. The Dubuque mill of Ingram, Kennedy & Day, now consuming upwards of 70,000 feet of logs per

[96] Contract between Ingram, Kennedy & Company and Burch, Babcock & Company, June 8, 1878.
[97] D. M. Dulany, Sr., to Ingram, June 5, 1878; Dulany to Ingram, Kennedy & Company, June 19, 1878.

day, obtained the greater portion of its sawlogs at Stillwater, Minnesota, on the St. Croix River. As in the valley of the Chippewa, the St. Croix was too low to allow the usual number of logs to reach the Stillwater booms, with the result that Major Day had no choice but to shut down the mill before the first of September. Conditions in the streams had to be most unfavorable for this to happen.

With so large a demand for lumber and so few logs available, it made for a frustrating season. Had lumber prices been higher the frustrations would have been even greater, but such was not the case. In fact, Dulany & McVeigh proposed a trade in August at figures slightly lower than their April offer, although the late summer proposition also included an advance of $10,000. But again no agreement could be reached, agent Potter explaining that "they do not want to advance money unless they get the lumber, or the money, this fall, and we are not *sure* of giving them either."[98]

For a time even Major Day seemed to be running short of optimism, concluding that the so-called hard times of former years had been "glorious" when viewed in terms of the present situation. After observing conditions in a trip across central Iowa, he advised Ingram that the "only class of people that now thrive in this section is the money shark." He said that the rate of interest on forced loans to save foreclosures and executions was generally 3 per cent per month.[99] In spite of what his friend Benton might charge, the farmers obviously faced real problems of their own.

As for the immediate future, Day thought there was no purpose in giving concessions on lumber because most dealers seemed to believe that prices would fall even lower. Further complicating problems, the area of northwestern Iowa, Nebraska and the Dakota Territory continued to be plagued by the grasshopper. The Major did note, however, that "Every paper and even [every] individual seems committed to the policy of silence in order that a hue & cry may not be raised that will retard emigration."[100] He personally held to the hope that a large number

[98] Potter to Ingram, Kennedy & Company, August 20 and August 22, 1878.
[99] Day to Ingram, November 24, 1878.
[100] Day to Ingram, September 15, 1878.

of unsuspecting homesteaders would settle on lands far to the west of the Missouri River, where they would serve as a buffer, enabling "the hopper to find food nearer home."

But the times were not so bad nor the prospects so gloomy as to prevent the Major from spending $3000 on some new gang-saw machinery for the Dubuque mill.[101] Furthermore, by some manner or means, Ingram, Kennedy & Day managed to sell 1,200,000 feet of lumber during the month of October. The lesson was becoming increasingly clear, as Charlie Horton had reminded Ingram earlier in the season, "there is a great demand for cheap lumber. . . ."[102]

Snow returned to the pineries in the winter of 1878–1879, and logging activities proceeded with greater ease and less expense than had been the case in recent seasons. In the spring, along the Chippewa, five Eau Claire firms co-operated in the log drives: Ingram, Kennedy & Company, Daniel Shaw Lumber Company, Badger State Lumber Company, Northwestern Lumber Company, and the Meridean Lumber Company. Their combined crews guided nearly 100 million feet of logs down to the improvements at Eau Claire. The Ingram, Kennedy & Company portion totalled some 35 million feet, and although the costs of driving varied according to the distances involved, the average rate was about fifteen cents per thousand. Unlike the previous two years, there was no lack of lumber in 1879. Ingram, Kennedy & Company marketed nearly 40 million feet during the season, sales amounting to more than $380,000. In addition, company agents also sold another 10 million contracted from producers who chose not to market their own lumber.

The only complaint common to lumber producers in 1879 was that the customers of Major Day had guessed correctly about prices when they postponed purchases in the fall of 1878. The average price for Ingram, Kennedy & Company mill-run lumber, not including culls, fell to barely $10 per thousand feet. Lumbermen had to look back through their records a good many years to find figures that were any lower. The decline, which Ingram presumed had at last reached its lowest point, began back in 1874. According to the company's financial records, the average

[101] Day to Ingram, October 22, 1878.
[102] Horton & Hamilton to Ingram, Kennedy & Company, May 8, 1878.

sale price to wholesalers in 1873 had been $13.43 per thousand board feet; in 1874, $11.93; in 1875, $12.34; 1876, $11.53; 1877, $11.15; 1878, $12.09; and 1879, $10.09, amounting to a decrease of $3.34 over the seven seasons.

Yet prices and profits were two separate items, and while the price on lumber had fallen considerably, costs had also declined. For example, 1879 was the first season in which the expenses of running the lumber from mill to market averaged less than $1 per thousand feet. That season Ingram, Kennedy & Company rafting costs averaged ninety-nine cents, whereas in 1873 these same costs had been computed at $1.60. Even at a ten-dollar selling price, there remained a margin on lumber in 1879.

There were, of course, limits beyond which costs could not be reduced, and operators like Ingram hoped that whatever it was that controlled the price of lumber, they would soon enjoy substantial improvement. But clearly the most significant consideration was that throughout the long depression, Ingram, Kennedy & Company continued to produce at a high rate. Although margins varied, and generally decreased, Ingram suffered no seasonal losses. His efficient operation produced cheap lumber and, good times or bad, there were customers for cheap lumber.

The decade of the 1870's produced further indication that successful lumbermen had to adjust the focus of their priorities to meet changing circumstances. Initially, in the late 1850's, it had been appropriate for Ingram to be preoccupied with activities associated with the actual production of lumber. At that early date there was an abundance of pine land readily accessible and relatively inexpensive, and there was no lack of customers for lumber of the quality sawed by his mill. In the 1870's greater attention was given to marketing activities, and the arrangements which Ingram constructed in this area of operations insured a large measure of security to Ingram, Kennedy & Company. In the 1880's new problems compelled Ingram to concern himself with the future supply of pine.

Thus, largely out of necessity, adjusting to the changing requirements, Ingram, Kennedy & Company evolved into a fully integrated operation of a more modern type, efficient in its techniques of production and secure in its sources of supply and arrangements for marketing. By 1880, however, Ingram was considerably more than a sawmiller. Indeed, it was his strength in marketing that brought him to the attention of Frederick Weyerhaeuser and the other members of the Mississippi River Logging Company. Ingram was too successful to be ignored.

CHAPTER VI

Local Victory
and Larger Defeat

ALTHOUGH THERE WERE SIMILARITIES between the
years following the Panic of 1873 and the post-panic years of
late 1850's, for Ingram, Kennedy & Company there were also
some important differences. One of the most important of these
had to do with the availability of credit. Better known and
established, the company enjoyed opportunities to obtain credit
during the hard times of the 1870's that had not been available
to the young firm prior to the Civil War. It may well have been,
moreover, that credit was even more crucial to the partners in
the decade of the 1870's than had earlier been the case. In part
this would have been true because of their increasing need to
invest funds in timber lands. By making full use of the credit
available, Ingram bought land on time while continuing to
direct surplus funds from the business operation into expanding
his production and marketing facilities. This was a very different situation than had existed ten to fifteen years previous, when
priorities for the company had to be so carefully drawn and
strictly followed. Fortunately, it was not then necessary to tie
up any large portion of their limited capital in land investment.
Just as fortunately, when such long-term commitments became
necessary, Ingram, Kennedy & Company had sufficient cash and
credit to adjust without undue hardship.

Still there were occasions when the availability of cash and
opportunities for credit were stretched to the limit. Had sawmilling machinery been more complex and costly, or had pine
land and stumpage been more expensive, firms such as Ingram,
Kennedy & Company would have soon been forced to resort to

the public sale of securities. As it was, although the organization at the managerial level gradually became more modern, for some time an expanding partnership continued to suffice in terms of ownership.

Buying pine land or stumpage on credit during the latter half of the nineteenth century was relatively painless because the rapid appreciation in the value of the timbered tracts more than financed such purchases, interest included. Indeed, the great fortunes that were built in pine lumbering were more the result of speculating in land than of producing lumber. Returns exceeding twenty times the original investment in less than twenty years were not uncommon.[1]

Although lumbermen like Ingram occasionally reaped some benefits from the inflation of land values, such benefits were always incidental, usually small, and limited by the fact that Ingram only purchased what he needed to supply his mills with sawlogs. It seemed never to occur to Ingram or his business partners that it might be wase to speculate in pine land. They were lumbermen in the strictest sense; when they bought land they bought it for the pine it contained and not for its potential. Even in the last decade of the century, when little pine at any price remained in the upper Mississippi valley and Ingram made some investments along the Washington coast, his purpose was not intentionally speculative.[2]

---

[1] One of the largest and probably the best known of those who profited from this considerable and continual increase in value of timber was Cornell University, whose trustees had wisely entered nearly half of their one-million-acre Morrill Act allotment in Wisconsin pine lands. Proceeds from the sale of these lands, with generous allowance for locating, maintaining and protecting them, and paying the "exorbitant" taxes, exceeded $5,000,000. Paul W. Gates, *The Wisconsin Pine Lands of Cornell University: A Study in Land Policy and Absentee Ownership* (Ithaca, New York: Cornell University Press, 1965 ed.). There are, of course, numerous illustrations of the extent and swiftness of the appreciation. As an example, Henry Sage, a New Yorker intimately connected with Cornell University, invested in some Michigan pine land near the Au Sable River after the Civil War at an original cost of $35,000. By 1882, the value of these same lands was $817,600. Anita Shafer Goodstein, *Biography of a Businessman: Henry W. Sage, 1814–1897* (Ithaca, New York: Cornell University Press, 1962), 139.

[2] In Washington he first purchased a mill site at South Bend on Willapa Bay. As the situation developed, he was not able to build a new empire, and the cedar, spruce, and fir he had purchased were sold to other lumbermen. But this had not been his original plan. Had he been a younger man he undoubtedly would have shifted the site of his operation from the valley of the Chippewa to the Pacific Northwest.

Not all producers of lumber were as reluctant as Ingram to purchase pine land. Frederick Weyerhaeuser, the Rock Island operator who had joined in the leasing of the Beef Slough Company rights in the fall of 1870, consistently purchased pine lands far in excess of his immediate needs. The choice tract of 840 acres of Chippewa pine land which he bought in 1870 was only a microscopic hint of what would follow. In the years ahead Weyerhaeuser would purchase for himself and for his interests thousands upon thousands of acres, and these investments are hardly explained by his grandson's wonderful observation that "Grandfather loved to buy timber."[3] Weyerhaeuser, like Ingram, was no mere speculator. He was, however, considerably more than a lumberman; he was manager of what became a vast national enterprise, and the interests he controlled and directed were so extensive that they demanded comparably large resources.

But these were later developments. In 1870, there were few in the Chippewa valley who had heard of Frederick Weyerhaeuser and fewer still who knew how to spell his name. Weyerhaeuser's subsequent importance has made it somewhat more difficult to recall with accuracy the nature of lumbering operations prior to 1870, at least in those areas in which he was to become involved. Along the Chippewa, for example, there has been a marked tendency to overlook the fact that there were a number of lumbermen like Ingram who had enjoyed relative success prior to Weyerhaeuser's arrival on the scene. If Frederick Weyerhaeuser was to have a dramatic influence on the history of the valley and on lumbering in general—and indeed he was—much had gone before.

In the course of the decade that began in 1870, Orrin Ingram was involved in many developments which, for the most part, were not subject to Weyerhaeuser's influence. Ingram supervised the expansion of all phases of his firm's operation in an effort to take the fullest possible advantage of the fair demand but relatively low price for lumber. He directed greater attention than previously to land purchases and to logging activities. He increased the milling capacity at Eau Claire both by obtaining new plants and improving the efficiency of the old. As Major Day had long requested, Ingram approved the addition of a sawmill

---

[3] Weyerhaeuser, *Trees and Men*, 14.

to the facilities at Dubuque; a number of additional retail outlets for Ingram, Kennedy & Day lumber were established across Iowa; and some new southeastern Minnesota yards were opened in connection with Ingram, Kennedy & Gill.

The railroad came and continued on, connecting Eau Claire with the rest of the world by new and exciting means. Although many local mills began to make use of this option in the delivery of their lumber, Ingram decided that for his firm's marketing arrangements, the advantages of water shipment remained uppermost. His decision was underlined by the construction of the *Clyde*, the iron-clad tow boat which would push Ingram, Kennedy & Company rafts from Reads Landing down the Mississippi for many seasons.

Ingram also learned that with success, and accompanying growth, he and his organization had to assume new responsibilities, sometimes distantly removed from the business of making and selling lumber. For example, it was expected that the more affluent citizens would contribute their influence, their wisdom, and at least some of their wealth to efforts to improve the health and welfare of the less fortunate members of the community, including such projects as the establishment of free public libraries, support of higher education, disaster relief, encouragement of temperance organizations, particularly the Young Men's Christian Association, and the like. To an inveterate church-goer like Ingram, these obligations were real.

Less easily accepted was a responsibility for representing the lumbering interests of Eau Claire, although in this matter there was little choice. An operator simply could not be large and successful and remain aloof from all that went on in hotel caucuses, in courts, and in the state legislative halls. Ingram never claimed to be particularly effective in the art of persuasion, but that in no way lessened his involvement; his unquestioned honesty and conscientiousness more than compensated for any forensic deficiencies. Orrin Ingram evoked trust, and he could not avoid a position of leadership in two critical and concurrent struggles.

The first and more important of these developed out of the earlier Beef Slough controversy, as Chippewa valley lumbermen continued their efforts to prevent, or at least to limit, the encroachments of the downriver interests led by Frederick Weyerhaeuser. This contest for supremacy on the Chippewa ended dramatically and decisively when the floods of 1880 swept a large portion of the logs belonging to the local operators past their sawmill booms and down into the hands of their rivals. The subsequent truce, negotiated without delay, permitted the Chippewa millmen to continue; but the rules of operation were no longer strictly their own.

Of less fundamental importance, although no less bitterly contested, was the so-called Dells controversy, a struggle between the lumbermen of Eau Claire and those of Chippewa Falls over proposed improvements at the lower Dells of the Chippewa. Not until the middle of the 1870's, after many years of frustration, did the Eau Claire millmen finally succeed in obtaining a franchise from the state legislature for the improvement of the river above their town so as to provide a large and secure reservoir for logs. The earliest settlers to the valley had realized, soon and painfully if not initially, the crucial importance of providing safe areas for holding logs. Even when most sawmills were powered by water, which was the case prior to the Civil War (the first mill of Dole, Ingram & Kennedy was only the second steam sawmill on the Chippewa), pioneer lumbermen appreciated that a splendid water-power site without adjacent storage was of no practical use. Indeed, the most serious handicap suffered by those operating mills at Chippewa Falls, with all of its magnificent water power, was the lack of any large area for safely holding logs. Just as clearly, one of the great natural advantages in the Eau Claire vicinity was Half Moon Lake, as Dole and Ingram acknowledged in their decision to locate on property between the river and the lake. No lumbermen anywhere enjoyed safer storage for their logs, though there were some problems associated with getting the logs out of the swift currents of the river and into the quiet waters of the lake.

Partly as a result of these problems, and partly because the lake could not be used by all of the local operators, the lumbermen of Eau Claire had long been interested in making certain improvements in the Chippewa mainstream at the Dells, a short distance above Eau Claire. The proposed improvements involved the placing of a dam across the Chippewa, the construction of booms and sorting pens, a cutoff canal bypassing a bend in the river thereby allowing that area to be used for log storage, and a canal or race between the improvements and Half Moon Lake.

The advantages of such improvements were obvious. But there were disadvantages for millmen located above the Dells. Simply put, changes that improved the river for one set of purposes might well impair it for another. In this instance, the purpose of the dam was to stop logs for the sawmills of Eau Claire. The mills above, however, were more concerned with the possible effects of such a dam on their lumber rafts. Was it possible to design and construct the improvements in such a manner that the rafting of lumber from points above could be continued without increased delay, danger, or expense? As early as October of 1861, the editor of the *Eau Claire Free Press* recognized this essential question when he offered mock assurance to his neighbors at the Falls that the proposed dam was "to be so ingeniously constructed that all lumber from above can be taken around the dam on wheelbarrows."[4] In truth, no improvements such as those contemplated by the Eau Claire operators could be placed in the river without increasing the difficulty of rafting lumber from Chippewa Falls. And, augmenting this very practical concern, was the animosity built of jealousy and competition between the neighboring communities. That the most natural inclination of the residents of Eau Claire and Chippewa Falls was not one of co-operation and consensus served only to lessen the possibilities for compromise.

Improvements at the Dells had been envisioned almost from the start of sawmilling activities at Eau Claire. Ingram had learned of such plans on his first visit to the valley in 1857 when Adin Randall, as a part of his sales promotion, mentioned them as if the matter had already been settled. In spite of Randall's apparent confidence, practical considerations forced the Eau Claire interests to delay the start of the Dells undertaking. One problem was to obtain a charter from the state legislature authorizing the improvements, a difficult requirement since Eau Claire and Chippewa Falls shared an assembly district until 1866. Thaddeus C. Pound, then partner in the Chippewa Falls lumbering firm of Pound, Halbert & Company, had been elected to represent the district in 1864, and this "mammoth, cultivated and representative brain of Chippewa County" made a political career of opposing every Eau Claire-sponsored improvement with

[4] *Eau Claire Free Press*, October 31, 1861.

all the power of his "mighty Websterian intellect."[5] Thad Pound was not going to sponsor or support the Eau Claire plans for the Dells in the Wisconsin Assembly.

But in 1866, a reorganization of the state assembly districts separated Eau Claire and Chippewa counties. In the election which followed, Fayette Allen of Pepin County received the honor of representing the new district which included Eau Claire, and with Joseph G. Thorp already in the Senate it appeared practicable to seek the desired charter for the Dells.[6] Thus, in early March of 1867, Ingram, in company with Daniel Shaw, Charles Gleason, George Buffington, and others, hurried off to Madison to assume membership in "the third house," the local editor observing that if they were defeated in their efforts, "it will not be by force of argument alone."[7] By some force or another, however, the Eau Claire interests suffered defeat by a substantial margin.

The failure of Ingram and his fellow advocates in the 1867 session of the legislature was largely attributed to a late start in their lobbying activities. They assumed, therefore, that success would come with greater and more timely efforts. But their opponents did likewise, with the result that while the Eau Claire pressures increased, the relative strength of the two groups remained substantially the same—and the Dells bill continued to go down to defeat.

The lobbying tactics adopted by both sides fell considerably short of propriety, even by the rather undemanding standards of the day. In February, 1871, when he was back in Madison to testify in behalf of the bill and to use to good advantage time not spent before the committee on labor and manufactures, Ingram wrote to Kennedy in Eau Claire admitting that he had lost much of his old enthusiasm and was becoming "thoroughly disgusted with [the] whole thing." He quickly assured his partner that their delegation from Eau Claire was "a very civil and quiet set," but unfortunately the same could not be said for those representing Chippewa Falls, whose hotel rooms were "full of whiskey and loafers night and day." Hopefully it would soon

[5] *Ibid.*, May 23, 1867.
[6] *Journal of the Wisconsin State Assembly*, 1867, pp. 754–755; Randall, *Chippewa Valley*, 140–141.
[7] *Eau Clarie Free Press*, March 7, 1867.

be over and Ingram thought they might carry it off, providing they could get it to "a square vote."[8]

Ingram was correct on one point; they did have the required number of votes to gain passage. Governor Lucius Fairchild, however, had sufficient doubts about there having been "a square vote" that he vetoed it.[9] Subsequent investigations failed to identify any individual members as having accepted bribes, but there was no question that the legislature as a body had been guilty of tolerating, if not condoning, illegal activities on the part of its own members and the lobbyists from the Chippewa valley. But the governor's veto, no doubt the appropriate response in 1871, had small effect in raising the standards of the conflict. Like Ingram, the other contestants may well have been as disgusted with the entire proceedings as was their governor. This did not alter the fact that logs continued to float down a "murmuring, undam'd-at-the-Dells Chippewa,"[10] and it did not lessen the importance of the improvements sought by the lumber interests of Eau Claire.

Forced to assume largely a negative position as concerned the proposed improvements at Eau Claire, the lumbermen of Chippewa Falls developed a positive program of their own. In the 1867 session of the legislature, they had succeeded in obtaining a franchise for the Eagle Rapids Flooding, Dam and Boom Company, which authorized construction of a dam at Eagle Rapids, about five miles above Chippewa Falls, for the purpose of enlarging boomage and storage facilities.[11] In one sense, the plans at Eagle Rapids were to Chippewa Falls what the Dells improvements were to Eau Claire—as the Dells proponents were quick to point out, accusing their upstream rivals of adopting "the dog and manger policy."[12]

It did not follow, however, that all Eau Claire lumbermen automatically opposed the efforts at Eagle Rapids. Indeed, Ingram was the first and only president of the Eagle Rapids Flood-

---

[8] Ingram to Kennedy, February 10, 1871.

[9] *Journal of the Wisconsin State Assembly*, 1871, p. 936; *Journal of the Wisconsin State Senate*, 1871, p. 745. For the text of the governor's veto, see Executive Message, *Journal of the Wisconsin State Assembly*, 1871, pp. 100–102.

[10] *Chippewa Herald*, April 22, 1871.

[11] *Private & Local Laws*, 1867, Chapter 328.

[12] *Eau Claire Free Press*, February 3, 1870.

ing, Dam and Boom Company, with Albert E. Pound, brother of Thad, serving as vice-president for the organization. It turned out to be a thankless and frustrating job for Ingram. His initial efforts centered on convincing his Eau Claire neighbors that it was truly in their interest to lend support to the Eagle Rapids Company; and after eventually succeeding in gaining the reluctant co-operation of a few, there was much difficulty in getting the improvement to work properly.

Originally the idea had been to stop the logs at the Eagle Rapids dam, sorting them so that logs intended for the big Union Lumber Company's mill at Chippewa Falls were directed into holding booms, allowing the balance to be released in a body and driven downstream to the Eau Claire facilities. The first dam was washed out before it had a fair opportunity to demonstrate any possible benefits, but the second structure lasted long enough to raise considerable question as to its value.

Ingram soon learned that improvements as extensive as those at Eagle Rapids tended to receive the blame for all that went on downstream, only some of which was deserved. The handling of the water of any sizable dam inevitably changes the character of the river below, and so it was with the Eagle Rapids improvements. In July, 1875, one of the early supporters of the project, Joseph G. Thorp, president of the Eau Claire Lumber Company, complained to Ingram that the running of lumber rafts on the lower Chippewa had been adversely affected as a result of improper handling of the dam at Eagle Rapids. Thorp observed that the periodic floods from Eagle Rapids were quickly followed by very low water and, as he described them, "flattened bars." He thought it likely that rafting conditions would be improved when the Eagle Rapids experts allowed nature to reassume control of the Chippewa's flow—and what Joseph Thorp thought was always important.[13] At about the same time, E. W. Culver, president of the Badger State Lumber Company, expressed dissatisfaction with the method of handling Eau Claire logs at Eagle Rapids. After reminding Ingram that his firm had paid nearly $2000 towards the construction of the improvements, Culver warned that unless Badger State received better treatment in the future, "we are done paying."[14] Central to Culver's complaint was the

[13] Thorp to Ingram, July 17, 1875.
[14] E. W. Culver to Ingram, August 16, 1875.

continued difficulty in securing what he considered a reasonable percentage of their own Badger State sawlogs, even on the planned floods from Eagle Rapids. A major part of the problem clearly involved the disinclination of the Union Lumber Company, formerly Pound, Halbert & Company, to work in true partnership with the Eau Claire millmen.

It was this failure which proved the greatest disappointment to Ingram as president of the Eagle Rapids Company. As he later recalled, by participating with the Chippewa Falls lumbermen at Eagle Rapids, "We hoped it would stop their opposition to the Dells dam, but they continued to oppose the proposition." Therefore, when floods again swept away the controversial improvements at Eagle Rapids, no third attempt was made to rebuild.[15]

But long before these works had demonstrated even their minimal utility, the Eau Claire operators realized that they would be no substitute for the improvements at the Dells. These millmen found themselves in quite a quandary. If they held to a position that legislative approval of the Dells project was bound to happen sooner or later, as most of them confidently believed, they still had to provide for their logs in the meantime. Yet to spend money on temporary improvements was not easy, since such improvements would become largely unnecessary with the completion of the project at the Dells. Lending additional urgency to their deliberations was the growing number of logs required to supply the ever-increasing demands of new and larger sawmilling facilities.

In the face of considerable pressure, the mill owners of West Eau Claire chose not to wait upon the whims of elected representatives in Madison. Early in 1868 they organized the Wilkins Island Improvement and Boom Company. Membership was limited to four neighboring firms: Ingram & Kennedy, Smith & Buffington, Daniel Shaw & Company, and Prescott, Burditt & Company, with George Buffington serving as president and Ingram as secretary for the new organization.[16] The total assessments, numbering three in all, amounted to $5358.75 for Ingram & Kennedy's seventy-five shares of Wilkins Island stock, which was doubtless an average for the four participating concerns.[17] The Wilkins Island Company built a network of booms and piers extending three-fourths of a mile along the river, capable of storing 20 million feet of logs. It was the most important and most successful of the various interim projects.

The works more than paid for themselves on two separate occasions during the summer of 1870, in periods of extremely low and extremely high water on the Chippewa. In the middle of the summer, many of the Eau Claire sawmills were forced to shut down for want of logs as a result of low water which prevented the driving of sawlogs from points above. For some weeks, only the west side members of the Wilkins Island Company had sufficient logs to remain in operation.

Although this situation graphically demonstrated the wisdom of the decision to go ahead with the plans for the Wilkins Island improvements, opponents of that project were reluctant to make such an admission. Still convinced that the money expended should have been saved for use at the Dells, where it would more generally benefit the lumbering interests of Eau Claire, they used the occasion of low water and no logs in midsummer 1870 to argue further on behalf of the construction of the Dells dam. Typical was the observation that the firms still producing lumber were "fortunate enough to have a stock of logs secured them by erecting booms along the river between this place and Chippewa Falls of sufficient capacity to hold their own logs, while lumbermen of less means, have been unable to secure booming privileges, and construct essential works; hence the necessity of a general log reservoir at the Dells."[18]

But Ingram and his Wilkins Island partners did not require the approval of others to appreciate their own current advantages, and this was especially true when the low water of midsummer was followed in late August by floods which ravaged the valley. In the course of a few violent days, improvements along the Chippewa were swept away, releasing the logs they had held. The booms of the Wilkins Island Company were among the very few on the entire river to remain intact.[19] The greatest benefits

[17] Wilkins Island Improvement and Boom Company stock certificates, February 3, 1869, in the Ingram Papers.

[18] *Eau Claire Free Press*, July 14, 1870.

[19] *Ibid.*, September 1, 1870.

associated with membership in the Wilkins Island Improvement and Boom Company, however, had less to do with extreme conditions on the river than with the handling of sawlogs under more normal circumstances from season to season.

These day-to-day advantages assumed increasing importance as the Dells charter bill repeatedly failed to gain legislative approval. Finally, in the 1875 session, the long-contested measure again passed both houses and nearly became law.[20] The governor did not exercise his power of veto; this time the Wisconsin Supreme Court invalidated the franchise, the justices maintaining that the bill as approved provided inadequate guarantees for the protection of the public interest. It was true that, as drafted, the bill's avowed purpose—the creation of a water works for the city—appeared to be of no greater import than the associated booms and log-sorting facilities.[21] In the 1876 session of the legislature, the authors of the Dells charter bill used greater discretion, stipulating that the primary purpose of the improvements was to provide a public water supply, and that such matters as arranging for the holding and the sorting of logs were purely incidental. The legislature voted favorably again and the court approved; the difference, of course, was one of words rather than intent.[22] As Ingram recalled years later, understandably confusing his nominatives: "We succeeded in getting a charter for a dam, or the city of Eau Claire did, for water-works ostensibly, but really to furnish a booming place to hold logs."[23] Thus, like the resolution of the Beef Slough controversy by means of a bill "to incorporate the Portage City Gas Light Company," the Dells question was settled ingloriously.[24]

---

[20] *Journal of the Wisconsin State Senate,* 1875, pp. 493–496; *Laws of Wisconsin,* 1875, Chapter 333.

[21] *Attorney General v. City of Eau Claire, et al.,* 37 Wis. 400 (1875).

[22] *Laws of Wisconsin,* 1876, Chapter 231; *State of Wisconsin v. City of Eau Claire, et al.,* 40 Wis. 535 (1877); Hurst, *Law and Economic Growth,* 172, 244, 253, 268–269.

[23] Ingram to Platt B. Walker, Jr., March 9, 1900.

[24] James Willard Hurst has correctly observed that the legislative response to both of these important questions shows "significant failure to fulfill the public interest in good legal process." Hurst, *Law and Economic Growth,* 269. But the participants apparently possessed sufficient cynicism to give small consideration to the injustice of either defeat or victory. Or perhaps there was simply too much to be done to permit pause and reflection.

It was to be expected that work on the Dells dam, so long delayed by the series of legislative, executive, and judicial decisions, would now be pushed to the limit of physical and financial resources. The city of Eau Claire then had neither the funds nor the fancy to undertake the establishment of its water-works at the Dells, regardless of the stipulations of the charter approved by the 1876 session of the legislature. Accordingly, the city turned over to the local lumbermen, organized for the purpose as the Eau Claire Dells Improvement Company, the task of constructing the improvements so long envisioned. Final agreement on the details of this transfer was reached on February 17, 1877, the city consenting to issue $95,000 worth of water-works bonds to aid in the cost of building the dam. This investment nearly matched the initial $100,000 capitalization of the Improvement Company. Even if the bonds had sold at face value, however, the combined amount would fall more than a hundred thousand dollars short of the total cost of the improvements.[25]

In spite of the large expense of the project, no one connected with the Eau Claire Dells Improvement Company would lose money. The log tolls alone soon provided enough to pay off the indebtedness. The real benefits, however, could not be measured on any ledger of the Dells Company as accurately as in the increased production of the sawmills of Eau Claire. During the peak lumbering years of the 1880's, the works at the Dells would handle 800 million feet of logs a season, and during the spring of these years, as much as 10 million feet a day.[26]

Orrin Ingram became the first president of the Eau Claire Dells Improvement Company, a responsibility that he accepted with considerable misgiving. It was a job with more than ordinary

[25] *Eau Claire Daily Telegram*, March 6, 1905. From the Eau Claire Dells Improvement Company accounts, dam construction cost $150,000; overflow rights, $18,000; construction of the one-and-one-half mile canal from the west end of the dam to Half Moon Lake, $11,776; the so-called Tainter gate and sluiceway, to carry downriver logs through the works, $10,162; the piers and booms used in scaling, sorting, and for storage purposes, $89,834; and $24,638 for the construction of the cut-off canal, for a total accounted cost of $304,411. See James Bruce Smith, "The Movements for Diversified Industry in Eau Claire, Wisconsin, 1879–1907: Boosterism and Urban Development Strategy in a Declining Lumber Town" (unpublished master's thesis, University of Wisconsin, 1967), 15.

[26] George Forrester, ed., *Historical and Biographical Album of the Chippewa Valley* (Chicago: A. Warner, Publisher, 1891–1892), 252–253.

opportunities for failure. Not only was it a very large undertaking, but so much had been expected for so many years from the improvements at the Dells that there was almost no possibility of avoiding many disappointments, in part because the supporters of the project had begun to believe their own exaggerated arguments as to its advantages. There were also, of course, many unforeseen problems which impeded progress on the works. With all of the difficulties, however, water began to fill the reservoir created by the new dam in March of 1878, barely a year after the arrangements for the start of construction had been completed.

No one could have been more pleased and relieved than president Ingram when it appeared that the Improvement Company's major facility was ready for use. The design for the dam of an estimated 4200-horsepower capacity was largely the work of H. H. Douglass of the Minneapolis Mill Company, who also served as a consultant to the U. S. Army Corps of Engineers detachment for the upper Mississippi region. The contract for construction of the dam had been awarded to McIntire Brothers, a New York firm with considerable experience in bridge, dam, and dock building. C. M. Buffington was hired to act as the Dells Company supervisor on the job. But engineers, contractors and superintendents could not relieve Ingram of final responsibility, and he spent many uneasy hours at the dam construction site where he seldom found all things progressing satisfactorily. He was bothered somewhat by the fact that James McIntire, president of McIntire Brothers, took a less personal interest in the progress of the project than Ingram considered appropriate. It also seemed to the president of the Dells Company that Buffington supervised from too great a distance. Ingram doubtless spent more time than necessary looking over shoulders or on his hands and knees, personally checking the details. The employees of Ingram & Kennedy had long since become accustomed to such visits from Ingram who, in truth, looked forward to opportunities to show his recognition of the importance of every task.

More crucial, and less enjoyable, were the problems Ingram encountered in finding money to pay for work on the dam. It was not an auspicious time to locate buyers for water-works bonds, especially from little-known points like Eau Claire, Wisconsin. Delos R. Moon, vice-president of the Northwestern Lum-

ber Company, spent some weeks in New York, Philadelphia, and
Boston trying to dispose of the bonds but without success. He
was followed by Charles R. Gleason, then city clerk, who stayed
two weeks in New York, spent $200 of Dells Company funds,
and sold not a bond. In the meantime, Ingram had been advancing money from the little bank in which he held an interest
with Dewitt Clark. When the amount advanced reached $40,000,
his banking partner indicated that he was unwilling for Clark
& Ingram to continue the practice. As Ingram later observed,
it was apparent that "the bonds must be sold at some price, and
that it was up to me as president of the Dells Improvement Company to make some disposition of them."[27]

Accordingly Ingram set off for New York, where, with persistence and a bit of good luck, he was able to make arrangements for sale of the bonds at ninety-five cents on the dollar
through the banking and brokerage firm of George Wm. Ballou
& Company. The terms agreed upon provided for an initial payment of $40,000 to cover the amount already borrowed from
Clark & Ingram, and $10,000 to $15,000 a month thereafter until
the balance was paid, a schedule not strictly adhered to due to
an almost complete absence of demand for the bonds.[28] But
Ingram's arrangements were, under the circumstances, surprisingly good. The Dells Company was thereafter able to maintain a
reasonable semblance of solvency.

For all of these difficulties and distractions of legislative
struggles and court battles, of victories and defeats, of subsequent
adjustments and new responsibilities, the lumber business went
on much as usual in the decade which began in 1870. The major
difference was simply that there was more of it; and because
there was more of it, the need for co-operation along the streams
and rivers of the valley increased substantially. This became
particularly apparent each spring with the commencement of
the log-driving operations.

Although pitched battles of the sort that featured the initial
Beef Slough Company efforts in the valley no longer occurred,
the sawlogs moved more slowly and at greater expense than was
necessary. The "outsiders," now organized as the Mississippi

[27] Ingram, *Autobiography*, 61.
[28] George Wm. Ballou & Company to Ingram, December 1, 1877.

River Logging Company, were firmly established in the Chippewa pineries. Under the leadership of Frederick Weyerhaeuser, these downriver operators were gaining control over a great area of valuable forest. Clearly their presence was no longer in doubt, and a certain amount of reluctant co-operation between resident millmen and the Mississippi River Logging Company took place every spring and summer. Actually, there was no practical alternative to this minimal co-operation, for neither group could afford the likely cost of interference and the resulting delays. But the co-operation that did occur was largely passive, and both parties must have appreciated the benefits which would accompany a genuine effort of alliance in those areas where their interests so obviously coincided. Certainly the log drive could not be conducted with any real efficiency until it became a joint effort under a single managing authority. The Eau Claire firms finally joined together in the later 1870's in the drives from the pineries to the new Dells improvements, but it was not until 1880 that the opposing interests of resident and downriver millmen agreed to work towards mutual advantage; and the agreement came about not because of the reasonableness of the participants, but from circumstances which forced a solution down stubborn craws.

Conditions throughout the winter of 1879–1880 had been generally favorable for logging, and although some of the spring drives proceeded downstream at a slower than usual rate, for the most part they had been clean. Frank Miller, in charge of the Hay Creek drive for Ingram, reported leaving only three IK (the common stamp for Ingram, Kennedy & Company) logs behind, "because they would not float."[29]

Charles Henry was Ingram's chief lieutenant on the drives, and it was Henry who had to make arrangements for co-ordination with others using the various streams feeding out of the forests down into the Chippewa. Although they had been working side by side for many seasons, co-operation still did not come easily between the crews of the Mississippi River Logging Company and those of valley firms such as Ingram, Kennedy & Company. In the spring of 1880, Henry became unusually aggravated with the activities of one Alex McDonald, foreman for

[29] Frank Miller to Ingram, May 11, 1880.

the Mississippi River Logging Company drive. The principal grievance in this instance involved the management of water within the system of flooding dams. It seemed that McDonald refused to lower the head of water behind some of his dams so that Henry's crews could proceed downstream with their logs. The slack water was too deep, or, as Henry explained to Ingram, "men can't do much with 14 foot poles in 18 or 20 feet of water." But try as he might, Henry could not convince McDonald that there were better methods of fulfilling their responsibilities. In disgust he wrote to Ingram, "If the President of the M. R. L. Co. and other parties interested in these logs were here on the ground they would agree with me when I say that he [McDonald] is no more fit to manage this drive, than I am to manage the affairs of this nation."[30] Neither Henry, his adversary McDonald, nor anyone else could possibly have known that 1880 was to be the last season for such problems in the valley of the Chippewa. Future drives might not always be perfectly clean, satisfactory, and successful, but they would definitely be co-operative affairs.

If Henry's reports from above Eau Claire displeased Ingram to a degree, he was encouraged by the news received from down-river markets. As in 1879, two separate crews conducted the selling and rafting operations for Ingram, Kennedy & Company on the Mississippi, one directed by Claire Chamberlin and the other by Allan Kennedy, a son of partner Donald. Prices opened considerably stronger in the spring than they had closed the previous fall, dealers above Dubuque offering as much as $13.50 for boards and strips and $11 for dimension lumber.[31]

The improvement was considerably less dramatic at Hannibal, where Allan Kennedy found that customers still "don't 'hitch' very well." Part of the difficulty was a reluctance among whole-salers to admit that conditions had indeed improved. Young Kennedy expressed regrets that they had made a trade with Hannibal parties late in the season of 1879 at the low figures of $9 and $11, observing, "Mr. Dubach can't get over it."[32]

Colonel Dulany, however, expressed what was for him un-usual optimism about the long range prospect for the lumber

[30] Charles H. Henry to Ingram, Kennedy & Company, May 5, 1880, in the Bartlett Papers, Eau Claire Public Library manuscripts collection.
[31] Dortland & Son to Ingram, Kennedy & Company, April 14, 1880.
[32] A. G. Kennedy to Ingram, Kennedy & Company, May 18, 1880.

business at Hannibal. In late May, he addressed a letter to Ingram requesting that lumber be allowed to come down to their market again, as much as could be spared, adding that if there were no customers waiting at his door, he would "pile it up for a rainy day," confident that sooner or later someone would buy.[33] The next day Dulany wrote again, this time expressing with directness the views that Ingram had been patiently waiting to hear from that direction: "I think it is to our mutual interests (yours & ours) to hold fast of this market here. it gives us both a market for our Stock, and we think we can handle all your surplus stock ourselves, and will always pay you the highest possible prices this market will justify."[34]

Concurrent with his growing interest in a binding arrangement with a firm such as Dulany & McVeigh in the Hannibal market, Ingram had been negotiating with Charlie Horton of Horton & Hamilton at Winona, also with the object of establishing a partnership outside the Ingram, Kennedy & Company organization. As a result, the Charles Horton Lumber Company took form in the early spring of 1880, although the membership was not widely published. (In fact, W. B. Judson, editor of the trade journal *Northwestern Lumberman,* wrote Ingram to inquire what he knew about the new company. Ingram knew a good deal, since he was vice-president and his assistant, Clarence Chamberlin, served as secretary-treasurer.) Horton was the president of the Charles Horton Lumber Company, but of course the important decisions were made in the Eau Claire offices of Ingram, Kennedy & Company. As a first step, Ingram directed Horton to begin looking over the rail line out of Winona for likely retail yard locations, as far west as there were customers.

Ingram's attention, lately given almost entirely over to marketing operations, abruptly shifted back to the Chippewa in the wake of a series of violent storms and heavy rains which swept across the upper Mississippi region in early June. Until that time, the drivers on the logging streams had been experiencing much trouble from too little water.[35] But by the end of the first week in June, streams were full and overflowing, and few log-

[33] Dulany & McVeigh to Ingram, May 25, 1880.
[34] Dulany & McVeigh to Ingram, May 26, 1880.
[35] William Fowlds to Ingram, Kennedy & Company, May 19, 1880.

driving dams proved capable of withstanding the combined effects of surging waters and battering logs.[36]

On June 9, the Mississippi was wild beyond belief. Work had to be suspended at Beef Slough, and Allan Kennedy noted a six-inch rise at Winona in twelve hours. The following day "high seas" swept Kennedy's leased raft boat, the *Hiram Price*, broadside into an island below the La Crosse railroad bridge, Kennedy reporting that rudders and steam were no match for the raging currents of the Mississippi.[37] The same day that the *Hiram Price* made its unscheduled landing, the ironclad *Clyde*, with Chamberlin aboard, was literally blown out of the river up on to an embankment, losing her raft of lumber in the process.[38]

But the most serious problems confronting Ingram were not on the Mississippi. The Chippewa was rapidly cresting at record heights, and the residents of Eau Claire braced themselves for the most disastrous flood in the history of their valley. With the light of new day on June 13 the worst was over, and it did not require the sleepless lumbermen much time to assess the extent of their damages. They had lost an entire season's supply of sawlogs; between 150 and 200 million feet had been swept over and around the Dells dam. These logs could now be of use only to their rivals, the Mississippi River millmen. As bleak as the situation at first appeared, however, Ingram realized that all was not darkness. For an initial consideration, the Dells dam remained intact. Also, damage to the sawmills had been surprisingly small. His own firm and some others could certainly obtain credit in almost any amount required. And yet, without logs, nothing else much mattered.

Insofar as they now controlled an entire season's supply of logs, Frederick Weyerhaeuser and his Mississippi River Logging Company associates held all the cards; and if a settlement was reached in the valley of the Chippewa, it would be on terms agreeable to Weyerhaeuser. So Ingram waited, receptive to say the least. He was not one to throw away a lifetime of work for the sake of past differences.

[36] John McLarity to Ingram, June 8, 1880.
[37] A. G. Kennedy to Ingram, Kennedy & Company, June 10, 1880.
[38] Chamberlin to Ingram, Kennedy & Company, June 10, 1880.

CHAPTER VII

Consolidation and Combination

THE DISASTROUS FLOODING along the Chippewa in the spring of 1880 brought about a revolution in the control and management of the pine resources of the valley. Frederick Weyerhaeuser dictated the terms of peace, and, not surprisingly, they embodied the very concessions that local millmen had long feared and fought against. Ingram and his associates soon learned with relief, however, that their capitulation was not the end of the world. On the contrary, for both resident and downriver operators the resulting adjustments proved largely beneficial from strictly a consideration of turning pine into profits; and, after all was said and done, that was the fundamental consideration. Although the settlement forced the operators of Eau Claire and Chippewa Falls to accept their downriver counterparts as more than just equals, for a time at least there was sufficient pine to share, and together they set themselves to the task of making more lumber and more money than most of them had imagined possible.

The substance of compromise between downriver and local interests was reached by mid-July, within a month after the waters of the Chippewa had begun to subside, though many details remained to be settled. It is no longer possible to say who assumed the initiative in arranging for the discussions leading toward settlement. It may well be that the Eau Claire and Chippewa Falls operators went to Roujet D. Marshall, who was then serving as a legal advisor to the Mississippi River Logging Company, and asked that he contact Weyerhaeuser on their behalf for the purpose of scheduling a time and place for meeting.[1]

---

[1] Hidy, *et al.*, *Timber and Men*, 73.

Some efforts at repairing the damage preceded negotiation of the larger problem. John Wither, Ingram's representative at Beef Slough, reported that the Mississippi River Logging Company employees were busily collecting logs that had run out of the works there and into the Mississippi. Three steamboats had been chartered and sixty men were employed in rounding up the strays before they became too scattered.[2]

Ingram soon became involved in similar efforts along the Chippewa. The day following the highest water at Eau Claire, he received a letter from E. K. Whitcomb, of the Chippewa Falls firm, Whitcomb & McLarren. Whitcomb expressed his hope that none would be discouraged as a result of the floods, counseling "let us go to work & repair our losses." He further suggested that they begin at once to hire all available teams to undertake the job of hauling back into the river the logs which the high waters had stranded along the length of the Chippewa mainstream, sometimes as much as three miles away from the normal channel. This proved an enormous task, since it involved the great majority of logs carried past the Dells improvements, a total exceeding one hundred million feet. But on July 15, Whitcomb thought that by setting up a few portable mills they ought to be able to saw a considerable amount of lumber, "besides giving employment to hundred[s] of poor laborers."[3] The situation confronting local operators, however, was far more serious than Whitcomb had yet admitted. Portable sawmills were indeed useful for the sawing of scattered pockets of escaped logs. But in the spring of 1880 nearly the entire harvest of sawlogs intended for valley producers had been swept below, an amount far beyond the capacities of such mills.

Few things were of less use to a sawmill operator than a log downstream. For the purposes of Ingram and the other Eau Claire millmen, the only logs which retained much value were those that had remained in the upper reaches of the valley, at points above Eau Claire, the largest portion of which bore the mark of the Mississippi River Logging Company. Thus the local operators had little bargaining power when they approached their showdown with Weyerhaeuser and associates.

---

[2] John F. Wither to Ingram, June 16, 1880.
[3] E. K. Whitcomb to Ingram, Kennedy & Company, June 15, 1880.

The relative weakness of their position did not prevent the Chippewa interests from taking steps to prepare themselves for the eventual confrontation with representatives of the Mississippi River Logging Company. On June 23, officers from eight of the valley's largest firms gathered in the Eau Claire offices of the Northwestern Lumber Company to discuss their common problems. They appointed a committee of Ingram, Delos R. Moon, and E. W. Culver, granting them full power to secure all of the logs stranded in the sloughs or aground on the banks of the Chippewa from Eau Claire above to the Little Falls Dam at Holcomb, where the Jump River and Main Creek joined the mainstream. The only instructions given Ingram, vice-president Moon of the Northwestern Lumber Company, and manager Culver of the Chippewa Lumber & Boom Company, were to put afloat all possible logs "in the most speedy and economical manner."[4]

The meeting between members of the Mississippi River Logging Company and the Chippewa operators took place in early July in the office of Chauncy Lamb at Clinton, Iowa. In assessing the subsequent discussions and decisions, the downriver millmen, particularly Weyerhaeuser, have received possibly more credit than they deserved for magnanimity. True, there seems to have been little or no expression of vindictiveness in the meetings. Yet it would have been entirely out of character for Weyerhaeuser, a genius of management, not to see beyond his present position of strength. He had not come to the valley of the Chippewa for the purpose of defeating its lumbermen. He had come for its pine, and he was anxious to get on with the business in the most efficient manner possible. It may be, as Roujet Marshall later recalled, that Orrin Ingram and his neighbors "were surprised" at the lack of a disposition on the part of their downriver opponents "to drive a hard bargain."[5] It seems far more likely, however, that after the early meetings Ingram simply felt relieved about the outlook and a bit anxious about the arrangements still to be negotiated. That Weyerhaeuser was intent upon forcing

---

[4] Agreement dated June 23, 1880, between representatives of the Daniel Shaw Lumber Company, French Lumber Company, Chippewa Lumber & Boom Company, Badger State Lumber Company, Valley Lumber Company, Northwestern Lumber Company, and Ingram, Kennedy & Company; copy to be found in the Ingram Papers.

[5] Marshall, *Autobiography*, 1:280.

co-operation rather than making new or keeping old enemies could hardly have surprised Ingram.

In truth, the Rock Island lumberman was not an unknown quantity to Ingram and the other resident operators in 1880. For ten years they had been working together. Every season Weyerhaeuser and Ingram had managed, with no more than ordinary difficulty, to arrange for the disposition of Ingram & Kennedy logs which arrived at the Beef Slough works by design or otherwise. During much of the summer, Ingram, Kennedy & Company had a full-time employee at Beef Slough who watched over the activities of the Mississippi River Logging Company crews, insuring that his employer's interests were not slighted. Even the drives of 1880, for all of the early upstream difficulties, later became a joint effort involving the Mississippi River Logging Company and the Chippewa interests, with the notable exception of the Chippewa Lumber & Boom Company of Chippewa Falls. The July of 1880 meeting in Clinton clearly was not the initial step leading to co-operative efforts in logging and driving activities in the pineries and along the streams of the Chippewa valley. An appreciation of the advantages of co-operation was not strictly a postdiluvian phenomenon.

In this respect lumbermen were similar to their counterparts in the profession of politics: they could be locked in what seemed a life-or-death struggle on one subject while simultaneously enjoying the closest business relations in another area of interest. This was no paradox. Successful businessmen had simply learned to separate the affairs of their business operation from personal feelings, though even the most successful forgot the lesson on occasion. Such occasions add interesting punctuation to their stories, but they are mere lapses.

Two separate problems received consideration at the Clinton meeting. The first was the recovery of the logs stranded along the banks of the Chippewa below Eau Claire. The second involved working out the details of a log exchange so that the Chippewa River mills might operate for the remainder of the 1880 season. Ingram and Weyerhaeuser, the acknowledged leaders of the two factions, assumed the responsibility for the negotiations, thereby insuring success or at the least an honest effort. The only remaining outsider operating in the valley was the large mill at Chippewa Falls—the Chippewa Lumber & Boom Company

—which, for purposes of its own, chose to use its strategic site to make things difficult for all below.

Ingram headed the committee responsible for recovering the logs. In combination with his earlier assignment of June 23, involving log recovery from Eau Claire above to the Little Falls Dam, he now assumed supervision for the entire effort along the Chippewa. Hauling these stranded logs back into the mainstream was a difficult and expensive task; on the lower river alone, the costs for recovery were estimated at $200,000.[6] Whenever possible, Ingram contracted with local farmers to haul the logs, often from their own land, to the water's edge and roll them in. In more remote areas, he dispatched crews to do the work. Payment for such a job was usually about seventy cents per thousand feet, with half of the amount paid in the middle of each month as the work progressed and the balance when all of the logs in a particular sector had been recovered. The millmen protected their own interests by inserting a provision in the hauling contracts which stipulated that if any logs were left behind, these would be scaled and their value charged against the account of the party doing the hauling. A suspense date was also included. Most of the contracts negotiated in mid-summer designated the first of October as the day by which the logs were to be afloat.[7]

As for the log exchange provisions, in general terms the arrangements provided that the Mississippi River Logging Company would pay the market price for logs belonging to the Chippewa mills that were then in Beef Slough booms or had been secured at points below along the Mississippi, or would be received as the hauling operations continued below Eau Claire. In return,

---

[6] Weyerhaeuser, *Trees and Men*, 11–12; Marshall, *Autobiography*, 1:278.

[7] These arrangements were so numerous and varied so slightly that Ingram and his lieutenants soon used pre-printed contract forms with appropriate blanks for names, dates, place, and payment details. See, for example, contracts between the Eau Claire Manufacturers' Driving Company and Daniel and Archie McClellan (July 13, 1880); James Hedrington (July 14, 1880); Frank Smith (July 16, 1880); Daniel McBeth (August 4, 1880); Donald McDougal and L. A. Blodgett (August 19, 1880); and James A. Mitchell (September 16, 1880). See also contracts between the Chippewa River Improvement and Log Driving Company and S. H. Taylor; Ole Hansen; John Gilmoure; and Marines Oleson, all dated July 17, 1880. These are all to be found in the Ingram Papers.

the Chippewa valley operators could purchase, also at the market price, logs which had safely remained at points above.

In effect, the agreement simply involved an exchange of logs, but there were some complications. Ingram specifically objected to Subdivision 7, a paragraph of the original contract between the individual firms on the Chippewa and the Mississippi River Logging Company which stipulated that the mills "shall not take faster than they receive without permission." The down-river millmen had intended this clause as a guarantee that the resident lumbermen would not take more than their share, their share amounting not to the number of logs they had put in above but the number of logs the Mississippi River Logging Company members received below. This was the essence of the exchange.

Although the purpose of Subdivision 7 was entirely clear and understandable, Ingram deemed it decidedly unfair to the mill owners of the Chippewa valley. They were in immediate need of sawlogs, while Mississippi River operators suffered no similar handicap. As he pointed out, if the intent of the provision were enforced, sawing operations in the Eau Claire mills would be slowed to the pace set by the crews just beginning the tedious work of hauling logs back to the river. Even Roujet Marshall's assurance—"You have now special permission to take the logs though the M. R. L. Co. *will not* get yours for some time"— failed to satisfy Ingram. Rather he sought evidence of good faith on the part of all participating parties, which he indicated should be provided by rescinding the objectionable clause. Lawyer and arbiter Marshall shortly arranged a compromise, which, he assured Ingram, "does away with your objection and preserves a check for them which they want."[8] In this small but important difference, Ingram assumed the role that he would maintain in future dealings between the local operators of the Chippewa and the Mississippi sawmillers. He became the watchdog for the old Chippewa alliance, fully appreciating the new arrangements for co-operation but at the same time realizing that the need for vigilance had not lessened.

Through the untiring efforts of Ingram, Weyerhaeuser, and many others, the recovery and exchange operations slowly began

---

[8] Roujet D. Marshall to Ingram, July 16, 1880.

to provide sawlogs to the waiting mills. It was apparent, however, that the Eau Claire lumbermen could not make up for
the time lost in the spring and early summer of 1880. All too
common were reports like the one received by Ingram, Kennedy
& Company from Guttenberg, Iowa, in which Chamberlin noted
that their German friends and customers were getting extremely
anxious about the receipt of lumber, adding, "I have done what
I could to quiet them without promising too much."[9]

In early August, Horton informed Ingram that they had pulled
every last board out at Winona and awaited the arrival of more
rafts with impatience. In addition, he confidently offered the
opinion that the Charles Horton Lumber Company alone could
handle all that the Eau Claire mills of Ingram, Kennedy & Company could provide. A week later, noting the continued increase
in demand at Winona, Horton predicted that lumber shipments
for August "from this place will probably be the largest on record." Unfortunately, the Charles Horton Lumber Company was
unable to claim its normal share of this large trade, primarily
because deliveries from Eau Claire early in the season had been
so wanting. As a result, there was insufficient seasoned lumber
on hand with which to take advantage of the large summer demand. As Horton explained, "our Lumber is to[o] heavy to get
in good car Loads," and he consequently had difficulty keeping
his own railroad yards at places like Goodhue, Lake Benton, and
Marshall, Minnesota, in good supply.[10]

Although production at the Eau Claire mills increased dramatically following the log exchange arrangements with the
Mississippi River Logging Company, the operators could not
saw lumber fast enough to satisfy their wholesale customers. By
the end of August, the crews at Horton's Winona yard were pulling and piling Ingram, Kennedy & Company lumber at an average rate of 125,000 feet per day. Even this rate of delivery could
not keep pace with demand. To make up the difference, Horton
arranged additional purchases from producers in Wausau, Wisconsin. Also in late summer, Horton oversaw the start of the
construction of a planing mill adjacent to their yard facilities at
Winona, an improvement which would allow the firm to take
advantage of the growing market for finished lumber. When

[9] Chamberlin to Ingram, Kennedy & Company, July 22, 1880.
[10] Horton to Ingram, August 2 and 7, 1880.

time permitted, he gave attention to the establishment of more retail yards along the railroads serving the region west of Winona, and in the fall of 1880 he completed plans for opening a new yard where the tracks crossed the Missouri River at distant Pierre, in Dakota Territory. Between drought and grasshoppers, the decade of the 1880's proved extremely difficult for most of the frontier farmers of the area; but these hard times failed to halt the Dakota land boom, and the Horton Lumber Company yard at Pierre was soon doing a large business in Ingram, Kennedy & Company lumber. By mid-November, the new yard was forwarding cash to Winona at a rate of $2000 per week, and Horton commented on the wisdom of "getting in there first."[11]

Horton was not the only wholesale dealer unhappy because of an insufficient stock on hand and slow supply in the summer of 1880. There simply was not enough lumber to go around. D. Dubach of Hannibal told Ingram of their "brisk" trade at that point, indicating their only complaint was that they had no lumber to sell. Asking if Ingram could not find some use for $15,000 then idling on his hands, Dubach acknowledged that he would be under "lasting obligation" if Ingram could arrange for the delivery of a raft or two. A few weeks later, Dubach repeated his offer, urging Ingram to "Draw when you wish . . . Don't be bashful."[12] But when Ingram subsequently took advantage of Dubach's proclaimed generosity he found—to no great surprise—that there were strings attached. Thus, in the middle of September, Dubach began to worry whether he would receive lumber equal to the amount of money loaned, and he complained that he had either been misunderstood or misused.[13]

For the company's downriver agents, the summer and fall of 1880 comprised a succession of effortless weeks in arranging for disposition of the Ingram, Kennedy & Company rafts. In August, Allan Kennedy observed that "it would be a pleasure to sell lumber on the present condition of the market," with every dealer wanting "a little Chippewa lumber." Although the younger Kennedy clearly wished that he had more lumber to sell under

[11] Horton to Ingram, Kennedy & Company, September 14, 1880; Horton to Ingram, November 16, 1880.
[12] D. Dubach & Company to Ingram, August 13 and August 24, 1880.
[13] A. G. Kennedy to Ingram, Kennedy & Company, September 14, 1880.

such favorable conditions, he was able to find "some concilation [*sic*] in seeing these parties who wouldn't buy when we had some, want it pretty bad."[14]

Business was also booming for Major Day at Dubuque. The spring floods which had caused so much difficulty along the Chippewa had assured the Ingram, Kennedy & Day sawmill of an unlimited supply of logs for the 1880 season. Not only was quantity no problem, but the quality of sawlogs was considerably above the average normally received at that mill. Many IK logs which had been intended for the Eau Claire mills ended up in Dubuque booms. Day was more than satisfied with the kind of lumber which came from those fine logs, remarking to Ingram: "The practical demonstration of their quality was constantly before my eyes in the yard."[15] Receiving almost no help from Eau Claire in the way of lumber, however, the mill in Dubuque could not produce enough to fill the orders coming to the Ingram, Kennedy & Day yard. As was the case with Horton at Winona, Major Day purchased some lumber at Wausau in an attempt to keep his assortment reasonably complete. For a brief time, the Major even discussed the possibility of "branching out" into the Wausau area, but the idea was soon forgotten. Still, that he even considered such distant expansion indicates something of the extent of the Major's energies, the demand for lumber, and the many distractions and responsibilities than devolving upon Ingram. One of the more obvious results of Ingram's many involvements was that Major Day enjoyed a much freer hand at Dubuque than previously. But Day was not entirely absorbed in lumber matters. On October 7 he married again; and, as proof of prosperity, took his Brooklyn bride to the Catskills for their honeymoon.[16] Unlike some of his hard-working associates, the Major was almost as adept at spending money as he was at making it.

The increasing independence enjoyed by Major Day was only one example of the changes taking place within the Ingram, Kennedy & Company family. Decentralization had to occur. Ingram

---

[14] A. G. Kennedy to Ingram, Kennedy & Company, August 17 and August 19, 1880.

[15] Day to Ingram, August 14, 1880.

[16] Day to Ingram, September 18 and October 13, 1880.

could no longer assert such a large degree of personal leadership, although the lesson was not learned easily and without considerable frustration. For a time he chafed under the burden of so many and varied demands, all of which seemed important. Decisions he had formerly insisted upon making were now being made elsewhere, in places like Dubuque, Winona, and Wabasha. He still had the continuing concerns associated with the Eau Claire mills of Ingram, Kennedy & Company, now larger than ever. Then there was the Dells Improvement Company for which, as president, he maintained the major responsibility. The problems involved in handling logs at Eagle Rapids remained a source of irritation. New and demanding assignments in combination with the Mississippi River Logging Company included supervision of log recovery, settling accounts, and negotiating complaints for damages. Finally, he was worried about his customers, at the moment so clearly in need of more lumber than his mills could provide. Orrin Ingram never forgot that he was in business primarily to sell lumber.

Within the company organization, the one who came closest to understanding the many problems facing Ingram was Clarence Chamberlin. The secretary should have been at his Eau Claire desk, but for the 1880 season George Potter stayed behind to assist Ingram in the office while Chamberlin took another turn as agent on the Mississippi. Chamberlin personally would have much preferred Eau Claire duty to being baked by the sun and devoured by mosquitoes, an unavoidable part of the rafting life. Ordinarily not given to complaint, Chamberlin nevertheless wrote Ingram from La Crosse, carefully inserting mention of the poor use then being made of his own abilities:[17]

> I can appreciate somewhat your situation. the constant demands upon you for your time and means. the constant watchfulness and ability required to meet all these demands at all times. But the judgment and ability that has so long characterized your business life will not certainly forsake you now. I wish there was more I could do to assist you. . . .

Ingram, too, wished that Chamberlin were closer to the Eau Claire office, but no one enjoyed the rafting phase of the operations and Potter was a partner, even if a minor one. In the

[17] Chamberlin to Ingram, September 18, 1880.

future, however, Chamberlin would not be absent from his desk
for any length of time, at least not for the purpose of arranging
for raft sales. As soon as rafting ended that fall, he was directed
to Winona, where he began to work over the books of the Charles
Horton Lumber Company.[18]

Given the continued demand for lumber, Ingram naturally
hoped that the rafting season of 1880 could be prolonged beyond
its normal end in early November. He hated to lose so much
business, and also hated to disappoint so many customers. But
instead of a longer rafting season, winter came unusually early
in 1880. Chamberlin reported ice on the raft the morning of
October 18, and two days later Horton informed Kennedy that
they had been unable to ship any lumber to points west of New
Ulm during the week previous because the railroad was "blocked
up with Snow."[19] Frequent blizzards and a lack of cars would
cause continuing difficulties for the Winona dealers throughout
the long winter.

But the winter of 1880–1881 was important for reasons other
than cold temperatures and heavy snows. There were many
problems to be dealt with during those months, for Ingram the
most pressing being to reorganize his firm. For this purpose,
he had dispatched Potter on a fact-finding tour in October,
gathering information of all kinds to be used in a review of the
condition of his many business investments. At Dubuque, for
example, during the Major's honeymoon absence, Potter learned
from bookkeeper C. J. Lesure that it would be surprising, indeed
disappointing, if the Dubuque yard did not show a profit of at
least $30,000 for the year, although the exact amount depended
on many figures yet unknown.[20]

Reacting to so many involvements and demands, Ingram had
begun to give serious consideration to the possibility of selling
his interest in Ingram, Kennedy & Day. This would have been
unwise, at least from the standpoint of reducing the burden of
his many activities and interests. In truth, thanks largely to
Major Day, Dubuque was no longer much of a bother. Thus the

---

[18] Chamberlin to Ingram, December 4, 1880.
[19] Chamberlin to Ingram, Kennedy & Company, October 18, 1880; Horton to
Kennedy, October 20, 1880.
[20] C. J. Lesure to Ingram, October 5, 1880.

more fundamental question had to do with Ingram's willingness to let the Major assume responsibility for the management of Ingram, Kennedy & Day. One thing was certain: Ingram could no longer make the business decisions at Dubuque. Either he agreed to give that authority to another, or he had to divest himself of that involvement. There were no alternatives.

Yet this was not an easy decision for Ingram to make. Up until that point, he had personally directed the organizations in which he had any large amount of money invested. That was no longer possible, but acceptance of the fact came slowly and with some misgivings. Only after much reflection did Ingram decide to retain his interest in Ingram, Kennedy & Day. Perhaps the presence of too many willing purchasers served to change his mind.

While Ingram pondered the question of his Dubuque involvement in late 1880, his relations with Major Day stretched thin. One of the major sources of friction had been over the matter of selecting a bookkeeper to replace J. W. Babcock in the Dubuque yard. The Major argued that it should be his prerogative to appoint Babcock's successor, but Ingram disagreed, insisting that someone of his own choice ought to be in the office there to look after the very considerable investment of the Eau Claire partners. It was not a question of honesty or integrity but of business, and Day finally consented. Accordingly, Charles Lesure of Eau Claire proceeded to Dubuque to assume charge of the books of Ingram, Kennedy & Day. The Major was informed of all details of the arrangement—with one important exception. Ingram, in keeping with his inclinations, offered Lesure a fractional part of the ownership, slightly less than a one-twentieth share, and Ingram assumed that Lesure's fraction would be taken out of Day's one-third interest. Just when or how Ingram planned on informing the Major of this levy is unclear. When Day accidentally learned that his new assistant was not just a bookkeeper, he was less than pleased; when he subsequently learned that Ingram intended that Lesure's fractional interest come out of his own one-third share, he firmly drew the line. Day stated that if Ingram wanted Lesure for a partner, Ingram could provide the opportunity for such a partnership. The profits which the partners later divided for the year 1880 totaled $43,807.44; the Major got his full one-third, or $14,602.48, and Lesure's

$2,688.55 was indeed subtracted from the Ingram, Kennedy & Company percentage.

Concurrent with these problems, Ingram had to deal with other organizational questions in the home office. Donald Kennedy, as he had long threatened, informed Ingram that he wanted to leave the company, leave Eau Claire, and begin a small business which would provide enough money with which to live and fewer worries to live with. No doubt a number of factors contributed to his decision, including his partner's many involvements outside of and far beyond the former limits of Ingram, Kennedy & Company's Eau Claire operation. Kennedy was probably more annoyed and worried than Ingram about the latter's involvements with Eagle Rapids, the Dells Improvement Company, Charles Horton at Winona, Major Day at Dubuque, and most recently with Frederick Weyerhaeuser and the Mississippi River Logging Company. It seems to have been this last development—what appeared to Kennedy as a willingness on the part of Ingram to serve another while neglecting their own interests—which proved decisive. With or without Weyerhaeuser, however, the two partners had grown apart. Kennedy simply could not do what Ingram was in the process of doing: accepting the change from old ways of owning and operating a business to new ways of organization, which sometimes involved a separation of ownership and management responsibilities. Furthermore, Kennedy was tired, confused, and certain that God intended him to enjoy at least some of the fruits of his labors. Ingram too was occasionally tired; but more often he was impatient to build and to improve, each year to produce and sell more lumber at greater margin. Like so many others, Orrin Ingram seemed committed to a quantitative assessment of progress, success and even happiness.

Ingram made no great effort to discourage the dissolution. Kennedy had become a bit of a brake. Things simply could not remain the same. Ingram appreciated that the existing partnership no longer served with any adequacy the rapidly expanding interests of Ingram, Kennedy & Company. For example, the recent organization of the Charles Horton Lumber Company had been, in form at least, an interim arrangement; Ingram looked forward to the day when growth could be managed within a stock company instead of having to establish new and separate partnerships for each expansion. From the beginning, Kennedy opposed such changes. In fact, as in the case of Tearse and Potter, he had objected to the use of one of the advantages of partnership—that of providing valuable employees a direct interest in the business. Though Ingram personally regretted ending his long association with Kennedy, at the same time he could hardly ignore the benefits which seemed likely to result.

So his old friend Donald Kennedy left the business and left Eau Claire. He did so on the best of terms, desiring only to begin anew in something small and satisfying, something over which he could retain full personal control from start to finish. This involved a move to Minneapolis and the formation with his son of Donald Kennedy & Son, whose most famous product was "The Columbia, acknowledged by thousands of users to be the best bed room commode in the World" (or so claimed the company letterhead). No doubt its design and construction were superb, but apparently its fame was not sufficiently widespread as to result in profits to the firm. In any event, Orrin Ingram and other Eau Claire acquaintances became accustomed to receiving requests for financial assistance from the Columbia's maker. Forever a friend, Donald Kennedy proved once again that his talents were largely mechanical.[21]

But in 1880, Kennedy had not been the only partner expressing reluctance regarding changes in organization. Major Day, already unhappy with matters relating to the appointment of Lesure, also viewed Ingram's plans with no great enthusiasm. Writing in late December, Day stated the nature of his concern to Ingram in clear terms:[22]

> With the changes now soon to take place in your firm or in any event likely to take place soon, new and unexpected combinations are likely to ensue, with sudden demise or otherwise ownership of stock is liable to change hands & I do not feel that it would be expedient for me to place so large a share of my capital where I might be left with no voice as to its control.

When the reorganization of Ingram, Kennedy & Day was completed about a year later, the Major's preferences received con-

---

[21] See, for example, Eugene Shaw to Ingram, June 27, 1893; Donald Kennedy & Son to Ingram, June 29, 1894; Ingram to William Irvine, June 30, 1894; and William Irvine to Donald Kennedy & Son, July 18, 1894.

[22] Day to Ingram, December 29, 1880.

siderable attention, with the result that the Dubuque operation continued on much the same basis as before.

Indeed, although the organizational change throughout the Ingram, Kennedy & Company family from partnership to stock company was a fundamental one over the long haul, initially any changes in day-to-day operations seemed slight. Ingram ceased to be quite the autocrat of seasons past, but the difference is probably better explained by the new demands then being made on his time rather than by any intent to democratize his business affairs. Certainly he had no concept of a modern corporation in mind. Shares of stock in the companies he controlled would not be made generally available. There would be no danger of any loss of owner control or other unsettling possibilities. The directors would be the major owners, for Ingram still believed that the business would best be managed by those who owned it.

By January 1, 1881, the arrangements had been completed for the sale of Kennedy's interests at Eau Claire and Dubuque, probably amounting to $150,000, to the Dulanys of Hannibal, Missouri. The Dulanys included, most prominently, brothers Daniel and William. Daniel M. Dulany, who because of his Kentucky ancestry and carriage was generally known as Colonel, was then sixty-five years old and married for the fourth time. This fourth and most durable spouse had borne him a daughter, but she was the Colonel's only child, and he listed the lack of a son among his life's major disappointments. William H. Dulany, two years younger than Daniel, had been more fortunate in the matter of progeny and had even named one of his three sons Daniel M. Dulany, Jr. The Colonel and William had been in the tobacco business prior to the Civil War, but in 1867 they formed a partnership with J. H. McVeigh and opened a wholesale lumber yard.[23]

Just as Ingram had long been interested in forming an association in Hannibal capable of handling large amounts of Eau Claire lumber, so the Dulanys had looked forward to the possibility of guaranteeing their yard a supply of the product from the Chippewa River valley. For some years past it had seemed likely that an arrangement with the Dulanys would be concluded,

---

[23] Howard L. Conrad, ed., *Encyclopedia of the History of Missouri* (New York: The Southern History Company, 1901), 2:333-336.

and perhaps only Kennedy's objections had delayed efforts in that direction. But it was also true that during most seasons there had not been much in the way of surplus lumber, at least to the amount necessary to justify any new major commitment. With Kennedy gone and more lumber on the way, plans could now go forward.

Since Colonel Dan was the recognized head of the Dulany clan, most of the negotiation leading up to the formation of the new organization had been between him and Ingram. Except on the subject of politics (both were ardent Democrats), the two were very different personalities. Unlike the stolid Ingram, the Colonel was a character to whom all events and people appeared in an extreme condition. Dan Dulany never lived an ordinary day; there were good days and bad days, and more of the latter than the former. Dan Dulany never met an ordinary person; there were friends and there were enemies, and considerably more of the latter than the former. Unlike the optimistic Ingram, the Colonel preferred to accept the bluest possible view of things. If business was good, the weather was bad; if the weather was good, then business was bad. In spite of their many differences, however, both the Colonel and Ingram were extremely capable businessmen. Their association would prove to be a very profitable one.

Along with the Dulanys, the other principal member of the new organization was already a business partner with Ingram at Winona, Minnesota. Charles Horton had been involved in lumbering for a long time. As a teenager in the early 1850's, he had worked in the Susquehanna valley pineries of Pennsylvania before coming west to Winona in 1856, the year before Ingram and Kennedy arrived at Eau Claire. At the depths of the depression in 1860, Horton was employed by a Winona sawmiller and received lumber in payment for his services that season. This lumber he somehow managed to sell and, with the cash received, he began a small sawmilling operation of his own. By the time he became associated with Ingram in 1880, the forty-five-year-old Horton had bought out a series of partners and was ready to do business on a larger scale.[24]

Thus, from a number of considerations, the time seemed ap-

---

[24] *American Lumbermen.*

propriate for expansion and consolidation. In early 1881, Ingram, Kennedy & Company, Dulany & McVeigh, and the Charles Horton Lumber Company gave way to the new Empire Lumber Company, capitalized at $800,000, with home office and sawmills in Eau Claire and wholesale yards in Winona and Hannibal. Orrin Ingram was president of the new organization; Colonel Dan Dulany, vice-president; Charles Horton, secretary; William H. Dulany, treasurer; and Clarence Chamberlin, assistant secretary. No change was then made or even considered in the organization of Ingram, Kennedy & Gill, and Charles Chamberlain continued to manage that partnership out of the yard in Wabasha, Minnesota.

Ingram acceded to Major Day's wishes that Ingram, Kennedy & Day be kept a separate operation, not so large that Day would have little to say regarding policies and decisions there. In fact, the more Ingram thought about the matter, the more logical it seemed for him to treat Day and Dubuque somewhat specially. Day produced his own lumber, accepted the responsibility for procuring his own logs, and, of course, had long managed the marketing activities of Ingram, Kennedy & Day with notable success. In effect, Ingram decided that Day had to be allowed an independent command although there would remain some definite limits on the extent of that independence. But beginning in 1881, the business responsibility at Dubuque belonged largely to the Major with Ingram assuming the role of an interested advisor. Ingram even agreed to Day's suggestion for a name for the new organization and, before the end of 1881, the Standard Lumber Company, capitalized at $500,000, officially announced its succession to Ingram, Kennedy & Day. The Major was president of Standard; Ingram, vice-president; George W. Dulany, the eldest son of William, was secretary; and Charles J. Lesure remained, changing his title from bookkeeper to treasurer.

The organizational changes which resulted in the formation of the Empire and Standard Lumber companies contributed to more efficient operation in the years ahead, years which would include the seasons of peak production and peak profits. The new arrangement was significantly stronger than the one it replaced, although in terms of relative strength the changes were not overly impressive. Such expansion and consolidation were common throughout the industry. The change was nevertheless important.

It was essential that Ingram, particularly in his dealings with Weyerhaeuser and the Mississippi River Logging Company, have a stable and secure organization as his base of support.

As important as the internal changes were to the Ingram-led business interests, they were hardly as crucial as the concurrent discussions between the Chippewa and Mississippi millmen, who were looking towards a more permanent solution to the problems which had long troubled them. Weyerhaeuser had indeed come to the aid of the resident operators following the floods of 1880, and the time had come to settle up. The Clinton meeting in July had merely provided the interim arrangements, sufficient to answer immediate problems and to sustain all parties for the balance of the season. It was understood, however, that negotiations between the groups would be resumed when the season was over, and the results of these talks would provide the basis for future operations in the pineries and on the log-driving streams of the Chippewa valley.

If there remained any local millmen yet unconvinced of the advantages of co-operation, the season of 1880 ought to have dispelled such doubts. Recovery from their near-disaster was nearly instantaneous in the early summer, since their only lack had been sawlogs and that deficiency was in large measure removed by the July agreement reached at Clinton. Although it may be stretching a point to conclude (as some historians have) that "the 'ruined' firms were soon in better condition financially than if the flood had never occurred,"[25] some producers did in fact enjoy a very successful year.

In the final year of the partnership, Ingram, Kennedy & Company, for example, sold more than 20 million feet of boards and strips at an average price per thousand of $13.33, and more than 9 million feet of dimension lumber at nearly $10.10, making an average price for mill run lumber of $12.36. This amounted to an increase of $2.25 per thousand feet over the 1879 season— the highest average price for any season since 1873. Including cull lumber and shingles, Ingram, Kennedy & Company's total sales for 1880 exceeded $410,000. Clearly, the business was large and prospering. The most significant figures concerning the Ingram, Kennedy & Company production during the 1880 sea-

[25] Hidy, et al., Timber and Men, 74.

son, however, was the recovery of 25,572,770 feet of IK logs below Eau Claire. Worth in a monetary sense somewhere around $200,000, the logs were literally irreplaceable. Their recovery once again indicated not only the importance but also the extent of the log exchange arrangements with the Mississippi River Logging Company.

In this connection, not all of the crews working along the banks and in the sloughs of the Chippewa completed their hauling operations in time for the logs to be exchanged and sawed during the 1880 season. But since the Clinton arrangements for exchange had been intended strictly for that season, the balance of logs recovered below their intended mills and not exchanged by the end of sawing in 1880 had to be provided for by individual contracts. On December 20, 1880, for example, Ingram signed a contract with the Mississippi River Logging Company for 7.5 million feet of sawlogs at $7 per thousand from the first IK logs to arrive at the Beef Slough works in the spring. If the advantages of payment seemed to belong to the downriver interests, recent experience and current negotiations convinced Ingram that continued good will was worth something.

Just a month earlier, at Chicago's Grand Pacific Hotel, a favorite gathering place for lumbermen, Ingram and other representatives from the Eau Claire mills met with members of the Mississippi River Logging Company to continue their discussions of the previous July. For all of their many differences, they had a very basic common interest—the efficient procurement of Chippewa sawlogs. With one exception, the resulting formal arrangements concerned that single subject. The November meeting in Chicago recommended the formation of an organization "for the purpose of uniting in one interest the purchasing, owning, driving, scaling, grading and distributing the several amounts of logs to each of the parties who shall subscribe and pay for them, in the best and cheapest manner."[26]

In large measure, the recommendation was merely a copy of the functions of the old Mississippi River Logging Company, but with one important difference: the new membership was to include firms from the Chippewa valley. Naturally, the crucial point of contention concerned the ratio at which the wealth of

[26] As quoted *ibid.,* 74.

the pineries would be shared between the two groups. Relative strength of the contesting factions was no doubt accurately reflected in the fractional division finally agreed upon, namely, two-fifths for the resident millmen and three-fifths for the Mississippi interests. This meant that although each member firm would receive logs in proportion to the amount of stock owned, the combined total of the Chippewa River members could not exceed about 40 per cent. A supplemental but significant result of these talks was a more-than-tacit acknowledgement by all participants that Frederick Weyerhaeuser was in command, evidenced by the fact that he alone obtained the personal authorization to arrange for purchases in behalf of the organization with no requirement to consult other members. It might fairly be said that if the agreement concluded in Chicago worked for the interests of all concerned, it worked best for the interests of Frederick Weyerhaeuser.

The arrangements became official on June 28, 1881, when Orrin Ingram, William A. Rust of the Eau Claire Lumber Company, and Delos R. Moon of the Northwestern Lumber Company organized the Chippewa Logging Company under the laws of Wisconsin. Ingram and Weyerhaeuser had negotiated the final details, including a stock division between the Eau Claire and downriver membership of 35 and 65 per cent respectively. Thus the famous "Chippewa pool" had its formal beginning. From that time forward the Chippewa Logging Company bought and sold, cut, drove, and distributed vritually all of the white pine of the Chippewa valley. Under the final terms only six Eau Claire firms joined with W. J. Young & Company, C. Lamb & Sons, the Hershey Lumber Company, Weyerhaeuser & Denkmann, Laird, Norton & Company, the Musser Lumber Company, and five other Mississippi River concerns in organizing the Chippewa Logging Company. But the six valley firms—the Empire Lumber Company, Daniel Shaw Lumber Company, Northwestern Lumber Company, Valley Lumber Company, Eau Claire Lumber Company, and Badger State Lumber Company—produced the great majority of Eau Claire lumber; and whether officially members or not, all parties engaged in logging activities within the area soon came to realize that a conflict of interests was no longer possible on the Chippewa.

The last important impediment to the unobstructed passage

of logs had already been removed by the recent purchase of the
Chippewa Lumber & Boom Company by the Weyerhaeuser-led
interests. No doubt in the hope of forcing downriver operators
to buy its holdings at the highest possible price, the management
of that Chippewa Falls concern for a number of seasons past had
made efforts to complicate log-driving efforts. In early 1881, its
strategic upriver advantages became all the more imposing when
the state legislature authorized the company to levy a toll of
fifteen cents on each thousand board feet of logs that passed
through its works.[27] For some time Weyerhaeuser had been interested in purchasing the huge Chippewa Falls firm, but felt
that the asking price had been excessive. With a prospect of a
toll on logs, however, he pressed for an early settlement. The
price demanded remained as high as before, but Weyerhaeuser
was now in a position to force his new Eau Claire "partners"
to share in the purchase. Thus on March 1, 1881, for $275,000
in cash and $1,000,000 in bonds, the stockholders of the Missis-
sippi River Logging Company and their Eau Claire associates, in
the shared proportions of approximately 65 and 35 per cent,
purchased what was then perhaps the world's largest sawmill
under one roof.[28] Included in the purchase price were 100,000
acres of pine land, half of which had not been logged and which
contained upwards of 450 million feet of stumpage.[29] Although
the main reason for the purchase of the Chippewa Lumber &
Boom Company in 1881 had been for purposes other than investment, it did prove to be a highly profitable property. Ingram
later estimated that the total dividends to shareholders ran between 5 and 6 million dollars, making his own long-term duties
as vice-president of that organization considerably less than bur-
densome.[30]

By the end of 1881, a new order clearly prevailed in the valley
of the Chippewa. Ingram was, of course, quite aware of all that
had happened, and his awareness was reinforced by the assessments which he was subsequently called upon to pay. For example, the Empire Lumber Company owned 85½ shares of

[27] *Laws of Wisconsin*, 1881, p. 319.

[28] See the *Chippewa Herald*, April 8, 1881.

[29] Marshall, *Autobiography*, 1: 284; Norton, *Mississippi River Logging Company*,
56; Hidy, *et al.*, *Timber and Men*, 74.

[30] Ingram, *Autobiography*, 54.

Chippewa Logging Company stock, on which there was an 1881 assessment of $100 per share, or $8550. They also held 1,260 shares of the Chippewa Lumber & Boom Company stock, on which there was a $22.50 assessment per share, amounting to over $28,200. Finally, the assessment on the Chippewa River Improvement & Log Driving Company stock, of which Empire owned 111 shares, totalled $8,538.45. The combined assessments for the related involvements came to $45,321.73 for the 1881 season, although Empire reduced its own cost by $7533.62, passing that amount along to the Standard Lumber Company of Dubuque, ostensibly in exchange for one-sixth of the benefits. No complaint was heard from Major Day. He could well afford to pay the bill, since the profits for the successor organization to Ingram, Kennedy & Day amounted to more than $80,000 for 1881.[31]

It was Weyerhaeuser who made the important lumbering decisions in the Chippewa valley following the spring floods of 1880. If Ingram remained cautious, alert, and aware, it was now strictly in the capacity of an advisor, hopeful that he might influence Weyerhaeuser. Weyerhaeuser needed associates such as Ingram who would serve his interests by serving their own, and Ingram well understood the nature of the new relationship. In a sense he had become a member of the loyal opposition who could maneuver only within the limits of the system. Years later he contended: "Had we not joined with the Mississippi men in the organization of the Chippewa Logging Company, neither Chippewa Falls nor Eau Claire would have amounted to very much."[32] This was strictly a retrospective observation. In 1880 and 1881, uncertainties abounded.

Though lacking viable alternatives, the local operators did not welcome with any enthusiasm the role of subordination. None had gone into business with a view towards eventually accepting some limitation on their methods of operation. On the contrary, the motivation for attempting to establish separate businesses normally involved a desire for independence. Suddenly such a desire was no longer practicable. Through circumstances beyond their control they found themselves in a situation similar

---

[31] D. M. Dulany, Sr., to Ingram, January 4, 1882.
[32] Ingram, *Autobiography*, 54.

to that of Ingram, their spokesman, who, in spite of the general success he had enjoyed since coming to the Chippewa valley nearly a quarter of a century ago, was now only partially independent at best. There was a certain precariousness involved in trying to maintain a modicum of former independence while working in a role definitely subordinate to Weyerhaeuser.

Thereafter the Chippewa millmen had to derive satisfaction mainly from the amount of money they made. In this respect they would have few grounds for dissatisfaction. For most of them, and perhaps especially for Orrin Ingram, the benefits of co-operation and subordination more than compensated, particularly in the area of timber purchases. Even before the Civil War, Ingram had objected to investing money in pine land; that sort of expenditure did not come easily to him. Under the new arrangements, the Empire Lumber Company was assured a supply of sawlogs by the Chippewa Logging Company for as long as pine remained in the valley. As depletion of the pine forest advanced, and the price of standing pine increased, co-operation in these matters became increasingly advantageous. Had competition continued to be a factor in the Chippewa pineries, the bidding on remaining timbered tracts would have proved damaging to operators with limited resources. Furthermore, although Ingram himself would have to delegate most of his marketing responsibilities, competition did remain a factor in the sale of lumber produced in the Empire sawmills at Eau Claire. Within the arrangements recently effected, the Empire directors could look ahead to such competition with confidence. Here, of course, they had to be careful. If they did not realize it initially, lumbermen throughout the upper Mississippi valley soon learned that before they did anything that might displease Frederick Weyerhaeuser, there were many consequences to consider.[33]

---

[33] Indeed, historians Hidy, Hill, and Nevins mislead when they assert: "Thus the coordination of activity on the Chippewa was a limited operation affecting one area, represented a voluntary union of varied interests on equal terms, and operated in a liberal manner." Hidy, *et al., Timber and Men,* 79.

CHAPTER VIII

Weyerhaeuser

IN THE EARLY 1880's Orrin Ingram probably knew as much as anyone about sawmilling and lumber markets in the upper Mississippi valley. In the decade ahead, his lumbering education was to be completed. Indeed, the president of Empire and director of the executive committee for the Chippewa Logging Company doubtless learned more about logging and about the topography of northwestern Wisconsin than he considered his due, especially at his age.

As noted previously, the process of change within the lumber industry through the years is, to a large extent, well illustrated by Ingram's own changing priorities. Following his arrival in the valley of the Chippewa, his immediate concerns had focused primarily on the sawmilling facilities, that is, the production phase of the operation. Pine land was then plentiful and stumpage relatively inexpensive. Customers from Glen Haven, Wisconsin, to Hannibal, Missouri, purchased the product of the mills, and if the trades were not always highly profitable, occasions were rare when rafts went unsold. In the 1870's, with mills producing increasing amounts of lumber and with continued access to land and availability of logs, Ingram paid greater attention to the problems associated with marketing.

Ingram had proved himself more than a sawmiller, and also more than a logger or wholesale dealer. He was a businessman who happened to be involved in lumbering. The general result of his business activities up to the early 1880's was a company by the name of Empire which produced fair lumber efficiently to the amount necessary for the supply of its own wholesale yards. With pine land rapidly decreasing in extent, and increasing in price, it was only natural that Ingram's future activities should

210

concentrate on matters relating to the supply of logs for the mills. Involvement with Frederick Weyerhaeuser and the Chippewa Logging Company did not alter the direction of this basic interest, although it doubtless altered the nature of that concern.

Changing priorities and new problems, however, failed to slow production or reduce profits. Throughout the decade of the 1880's and into the early 1890's lumbermen prospered as never before. The prosperity encouraged expansion and still greater expansion until the occurrence of another depression, a depression that struck with a severity unprecedented in the lumber markets of the upper Mississippi valley.

Yet the Panic of 1893 was not the fundamental cause for the end of the peak years of lumber production and profits. There was a practical limit to the expansion of lumbering activities. The logger was once again proving the myth of forest inexhaustibility, and by the early 1890's lumbermen from the Chippewa valley and neighboring districts, fast reaching the limits of the Lake States pineries, were beginning to look elsewhere at unexploited forests more than a thousand miles to the west.

In a sense, the new arrangements for logging and driving in the Chippewa valley proved too efficient, at least from the consideration of permanence of operation. Although figures alone can tell but a portion of the story, they do serve to indicate something of the magnitude of the expansion of operations.[1] During the 1868 season, for example, the first year in which the "outsiders" made use of Beef Slough, a total of not more than 15 million board feet of logs reached those quiet waters. By July, 1888, as many as 11 million feet of logs left the mouth of the Slough in rafts *each working day*.[2] It must also be noted, however, that if cutting increased at a faster and faster rate, the lumber produced was not going to waste.

The same could not be said of the trees in the forest, but even logging practices gradually began to show evidence of reform as the value of standing pine increased. With the passage of seasons, loggers felled smaller and smaller trees and used more of

[1] See Appendix, Table I, p. 291 below.

[2] *Mississippi Valley Lumberman & Manufacturer*, July 6, 1888; Walter A. Blair, *A Raft Pilot's Log* (Cleveland: Arthur H. Clark Company, 1930), 48–53.

[3] Compiled by the U.S. Army Corps of Engineers, *H.R. Engineer's Report*, 1st Session, 57th Congress, Vol. 14, p. 2328.

the fallen tree than had previously been their custom. By the 1890's, sawyers took down trees that their axe-equipped predecessors of the 1870's would have scorned, either because they were too small or in some manner imperfect.[4] Still, by any recent standards, a surprisingly large amount of the pine was left standing or lying on the ground near the stump after the loggers moved on. One of Ingram's logging foremen, Bill Miller, reported during the 1880–1881 winter that he was quite satisfied with the progress of the work and the quality of the logs, adding, "of course we leave a good deal in the woods that we think you don't want."[5] Circumstances slowly forced the lumbermen to lower their standards of sawlog quality; yet, in 1893, the chief of the United States Division of Forestry estimated that "hardly more than thirty to forty per cent of the wood in the trees that are cut down reaches the market. Sixty to seventy per cent, rarely less, is left in the woods unused."[6]

But the mere fact that wood was undeniably wasted did not make the operation uneconomical. On the contrary, to the nineteenth-century logger it was uneconomical to be concerned with anything other than a near-perfect log. He could not afford to struggle with a tree, even after it had been felled, if that tree were diseased, damaged, or imperfect to any extent. The price of logs simply could not bear the costly considerations of conservation. This was true despite the fact that these forgotten trees and trimmings formed massive accumulations of slash which were the prominent cause of the most destructive forest fires. A fire burning through a clean forest might kill great numbers of trees, but if given prompt attention, much of the

[4] Henry C. Putnam of Eau Claire, well-informed about Chippewa pine land, reported that in 1880 loggers cut trees with a diameter of twelve inches, twenty-four feet up from the stump, and that by the early 1890's this minimum size had shrunk to seven or eight inches. See Putnam, "The Forests of Wisconsin," State of Wisconsin Horticultural Society, *Transactions*, 1893 (Madison: Published by the State), 184. The mention in the text of saw and ax is actually a reference to an interesting technological lag in logging operations. Tradition was of some special importance in the woods, and apparently this delayed the adoption of the saw as the basic tool of logging until the late 1870's. *Mississippi Valley Lumberman and Manufacturer*, November 10, 1876; *Northwestern Lumberman*, December 9, 1876.

[5] William Miller to Ingram, Kennedy & Company, December 30, 1880.

[6] Bernard E. Fernow, "Difficulties in the Way of Rational Forest Management by Lumbermen," State of Wisconsin Horticultural Society, *Transactions*, 1893 (Madison: Published by the State), 190.

wood could still be used. Slash not only made a tinder box of the forest, but slash-fed fires left behind little more than blackened stumps. Nevertheless, the waste continued because cheap lumber was in demand and because the extravagant methods of logging contributed to its cheapness. It may be argued that this was not true economy; but to the producer of that period, it was the only way to operate. To conclude otherwise, one would be forced to demonstrate by some method or other that these lumbermen did not know their business, a demonstration which would be difficult at best.[7]

As noted, a particular advantage to the Empire Lumber Company resulting from its membership in "the pool" had to do with the security of log supply. This did not mean, however, that beginning in 1882 Ingram could ignore the logging phase of operations to any extent. In fact, he made his first lengthy trips to the logging camps in the winter of 1881–1882, and his primary responsibility in the seasons to follow was involved with the cutting, grading, scaling, and pricing of logs. These new demands were directly related to the role he accepted within the new organizations, the Chippewa Logging Company and the Chippewa Lumber & Boom Company. But with or without the arrangements imposed on the valley by Weyerhaeuser, it seems probable that in the decade of the 1880's Ingram would have devoted an increasing amount of time to the subject of sawlog procurement. When lumbermen began to estimate the amount of standing pine, the end of the forest was clearly in sight; and at that point pine trees and sawlogs could no longer be taken for granted.

Yet Weyerhaeuser himself was a special and complex problem for Ingram to face. There were very few individuals that the president of Empire dealt with on less than equal terms, and "the man" was obviously one of them. As a result of this basic inequality, Ingram had to present his views cautiously, whatever the subject, especially since his views were often at variance with those of Weyerhaeuser.

Ingram continued to serve as the unofficial liaison between the old factions, residents and "ousiders." In some respects, this became an easier assignment with the passing years. The sawmill operators, more readily than the free-wheeling loggers, ac-

[7] Twining, "The Lumbering Industry in Perspective," 123.

cepted the loss of their decision-making power and the system in which they invested money without the possibility of asserting any real control. For the most part they made the transition with relative ease, assuming their former functions within the Weyerhaeuser organization or concentrating their efforts and attention on processing and marketing activities where some pretense of independence could still be maintained. Nevertheless, there were times when they recalled the old days and the old ways and wished again for what they had lost. So in a sense Ingram was more than a liaison between individuals; he was a liaison between periods.

Ingram represented the minority interest within the Chippewa Logging Company as surely as Weyerhaeuser represented the majority interest. But this was no true democracy, and Ingram seldom felt at liberty to make his points with the firmness and confidence which came naturally to Frederick Weyerhaeuser. Although the combination of resident and downriver interests in the Chippewa Logging Company resolved the differences and provided generally efficient co-operation in logging and driving operations in the valley, the decisions reached along the way were not always unanimous. The marriage had been one of convenience if not necessity, and differences remained much as before.

At the center of the continuing controversy was the pine log and the important season-to-season questions of how many logs to cut and what the price per thousand feet ought to be. In the early 1880's, Weyerhaeuser apparently gave considerable attention to the creation of an image of honesty and fairness. From the extent of his involvement, if for no other reason, this proved a most difficult task. As an example of the situation in which Weyerhaeuser operated, in December of 1885 the Chippewa Lumber & Boom Company of Chippewa Falls sought to purchase some pine land—land which happened to be largely owned by Weyerhaeuser. Since Weyerhaeuser was president of the Chippewa Lumber & Boom Company and, of course, a very major stockholder, he was doubly involved. In this instance, faced with so obvious a conflict of interests, Weyerhaeuser attempted to remove himself from the negotiations. William Irvine, mill manager for the Chippewa Lumber & Boom Company, wrote to Ingram, then vice-president of the company: "As Mr. Weyerhaeuser is interested in this timber he declines to have anything to say in regard to it, and suggests that I communicate with you in the premises."[8] On the surface such an acknowledgement might seem commendatory; but Ingram and Irvine both knew that Weyerhaeuser could not remove himself. His interests were too widespread, too large to be avoided, even if such avoidance were desired. A correct decision by a subordinate in such matters was one which agreed with Weyerhaeuser's own views on the subject. Anything less would likely result in the reopening of negotiations in order to reach a new decision, hopefully acceptable to the one whose opinion counted. Small wonder that Ingram and others found decision-making scarcely more than a formality.[9]

For many seasons Ingram served as director of the three-member executive committee of the Chippewa Logging Company, along with Weyerhaeuser and W. J. Young, who was later replaced by Henry Morford. The primary responsibility of this committee was to grade and price the logs put in at the various camps. Here too Weyerhaeuser was capable of making arbitrary changes if he decided, in consultation with himself, that such changes were in order.[10]

This dictatorial tendency was without doubt the source of greatest discontent among those associated with Weyerhaeuser. These attitudes are exemplified by a minor controversy in early 1890 which involved the disposition of the surplus funds of the Chippewa Lumber & Boom Company. Specifically, the small stockholders wanted the surplus put into dividend payments, an

[8] William Irvine to Ingram, December 30, 1885.

[9] Weyerhaeuser was quietly hated by many. As would be expected, dissatisfaction tended to increase as time went on; but public complaints were few. More commonly they were shared by close friends, as when Will Tearse wrote to Clarence Chamberlin on March 21, 1892, inquiring as to what action had been taken at a recent Chippewa Lumber & Boom Company meeting. Tearse thought he knew what had happened: "I suppose after Fred told you what he was going to do about it, you all went quietly out of the door, to take the train for home."
It should be noted, however, that Weyerhaeuser was simply the strongest among a group of men in which strong personalities were hardly the exception. Ingram, for example, was accustomed to dominate most of those with whom he associated, and this included many who must have resented his authority just as he resented Weyerhaeuser's. One-horse loggers would occasionally mutter angrily concerning their treatment at the hands of "Rule or Ruin Ingram," and business partners would complain of O.H.'s autocratic tendencies. Clearly, at times it must have been difficult to be a Christian businessman.

[10] See, for example, B. E. Reid to Ingram, June 11, 1897.

action which Weyerhaeuser opposed. Weyerhaeuser personally had no immediate need for the money, and preferred that the company invest its surplus funds in pine lands. If individual stockholders were particularly hard-pressed for cash, Weyerhaeuser proposed that the company should buy back their shares, again making use of surplus funds.

Major Day, one of the smaller shareholders, expressed disappointment in "the sense of honor of Mr. Weyerhaeuser and some of his associates" who somehow seemed unaware or unconcerned that buying out the interest "of needy and weary stockholders" with funds which should have been paid in dividends "was not making fair use of the powers entrusted to them." Furthermore, the Major questioned why the stockholders were not consulted about such crucial matters as the amount of logs to put in during a logging season. He knew, of course, that the stockholders were not consulted because Weyerhaeuser made those decisions for them.

This was no isolated instance. Major Day often sympathized with Ingram, noting that "in this matter, as well as in the purchase of logs, the minority interest such as you represent, is given but small account, and the relations existing between yourself and the other officers of the pool must become more and more strained and unpleasant." Day was not alone in accepting with difficulty his role within the new order of things. But at least he had the satisfaction of complaining to Ingram:[11]

> I do not believe in the one man power management of Mr. Weyerhaeuser. His business capacity and foresight may be of the highest order, and every move he makes may result in the greatest advantage to the company, notwithstanding that many of us at present are unable to see that probable result, but if these things were true, it is hardly becoming for self respecting, able, and competent business men to surrender all their rights and interests into the hands of one man, and that man inclined to act on decisions adverse to the will of the majority, and furthermore, a largely interested seller as well as a most liberal and profuse buyer. Mr. Weyerhaeuser may not be as much to blame, as the men who have for years consented to his dictation, and thus encouraged a natural bent to independent action to a point where he becomes the company.

[11] Day to Ingram, March 25, 1890.

Major Day was certainly correct about one thing: Weyerhaeuser was the company. Yet one has the distinct impression
that if such complaints had reached the ear of Mr. Weyerhaeuser,
his response would have been something to the effect, "If Major
Day is unhappy with the return on his investment, I will be
pleased to purchase his interest in the company." Seldom did such
complaints reach Weyerhaeuser, however, and for the most part
it was not easy for those involved to express serious displeasure
while turning such large profits.

Few sawmill owners, and certainly not Major Day, were required to make as large an adjustment within the new order of
things as Ingram. When the fifty-one-year-old president of Empire started off for the logging camps in the winter of 1881–1882,
he was, to a considerable extent, starting all over again. To one
who attached some importance to independence of action and the
fact that he had gained success largely through his own efforts and
abilities, there had to be periods of depression with the realization that, in truth, he was once again in another's employ. For
the most part, however, Ingram kept these things to himself.
He doubtless found some consolation in recalling that there had
really been no choice in the matter of accepting the arrangements
which Frederick Weyerhaeuser had offered. Indeed, the situation could have been far worse, and Ingram had always been one
to make the best of things.

To those who knew him, Ingram seemed no different than before. He managed his new responsibiilties with serious purpose
and in good spirit. Aside from his natural inclination to do business on the best possible terms, he was further motivated by the
fact that he could not expect Weyerhaeuser and his Mississippi
River associates, by now the owners of large quantities of Chippewa land and stumpage, to act in accordance with interests other
than their own. If Empire and the other resident firms of the
Chippewa Logging Company were to be victimized or in any
way to suffer discrimination in their dealings with the downriver
millmen, Ingram wanted to be certain that the facts of the situation were known to all. Or, viewed in more positive terms, if
vigilance and hard work could protect the interests of Empire
within the Chippewa Logging Company, Ingram would see to
it that those interests were protected.

It may well have been just this sort of dedication that Weyerhaeuser had counted upon to make the organization function effectively. In his own far-ranging "empire," Weyerhaeuser depended to an ever-increasing extent upon assistants such as Ingram. When local operators like the president of the Empire Lumber Company assumed the burden of managing the details of an organization like the Chippewa Logging Company for the purpose of protecting their interests, Weyerhaeuser's own interests would likewise be served. In other words, while he doubtless realized that few men could be trusted, he also knew that in their efforts "to keep Weyerhaeuser honest," it was likely that all dealings would be checked and double-checked.

It was not, of course, to please Weyerhaeuser that Ingram accepted the new responsibilities; it was strictly to serve Empire. Accordingly, on those few occasions when Ingram felt especially depressed, it was usually when his own associates in the Empire Lumber Company failed to appreciate, or even belittled, his efforts in their behalf. Ingram often found it impossible to satisfy Weyerhaeuser while attempting to do justice to the Empire interests; compromise became the closest thing to success. Furthermore, Ingram experienced some frustrating moments because he could not tell Weyerhaeuser exactly how he felt. Consequently his Empire subordinates ran the danger of receiving a double dosage of reaction when they persisted too long in their complaints. To Ingram, the work itself was bad enough; to be unappreciated was much worse. This seemed especially true with regard to the winter journeys into the woods. Ingram did not want sympathy, but he wanted it understood that he did not undertake trips through the pineries for pleasure.

There were measures which Ingram and Empire could take which afforded some protection in their dealings with the pool. Although the chief benefit derived from Chippewa Logging Company membership was security of log supply, each member firm receiving logs in proportion to its share of stock, this provision did not decrease the importance of logging activities for firms like Empire. On the contrary, as Ingram viewed the matter, logging became all the more crucial. At the crux of his consideration was the importance for Empire to put in as many logs as it expected to receive.

The Chippewa Logging Company owned pine land and stumpage and operated a number of its own logging camps. But it also purchased logs from the independent operations of member firms as well as from loggers operating entirely outside the pool. Officers of the Chippewa Logging Company, or more precisely the executive committee of which Ingram was the director, agreed on a total of logs to cut and to purchase each season. For example, in 1886 a total of slightly more than 475 million feet was received by member firms, nearly 340 million feet of which were allocated to the Mississippi River Logging Company member mills. The executive committee would also stipulate the price of pool logs, a most important and delicate determination. In this situation, the Empire could guarantee some self-protection by continuing to conduct its own logging operations. Again using the example of 1886, Empire's share amounted to nearly 36 million feet, and it was this amount which Ingram hoped to approximate in his own camps during the 1885–1886 logging season. Thus, no matter what price would be assigned to pool logs, Empire would come out even.

Beginning in 1882 this attempt to match the amount of logs allocated under the pool distribution arrangements was the consideration at the basis of the logging operations of the Empire Lumber Company. Even Colonel Dulany, who normally opposed timber purchase at any price and for any reason, recognized the importance of maintaining some degree of independence in this area. He agreed that "from year to year" they should secure sufficient pine land so as to be in a position to do a good portion of their own logging, which "would protect us against those large owners of stumpage . . . from having to pay them any price after awhile that they might ask for Pool Logs. . . ."[12]

In general the system worked well. In the first five years of operation the pool distributed a total of 2,268,024,260 feet of logs. Of that amount Empire received just slightly more than 165 million feet. In relation to the total distribution, the percentage directly benefiting Ingram seems small indeed. This is especially true when considering the extent of his responsibilities in the Chippewa Logging Company and the fact that his only compensation came from the profits of Empire. Such reflections, however, served little purpose since he could hardly get out. In a sense, the other small member firms of the Chippewa Logging

[12] D. M. Dulany, Sr., to Ingram, December 25, 1882.

Company were using Ingram as much or more than was Weyerhaeuser.[13]

Although Ingram made his first extensive trips to the logging
camps during the winter of 1881–1882 in connection with his
executive committee responsibilities, he already knew a good
deal about that phase of operations. For the most part, however, his interest in logging had been abstract and his knowledge
was second-hand, focusing on the more tangible aspects, such as
the amount of No. 1 and No. 2 logs banked by a particular camp
and the cost of the job per thousand feet. He knew, for example,
that it could be more difficult and more expensive to get hay
and provisions into the pineries than it was to get the logs out.
He also understood that success and costs of logging depended
in large measure on favorable weather conditions, specifically
low temperatures and sufficient, though not too much, snow.

But in all probability, Ingram had considered logging to be a
fairly simple business, just as the problems of manufacture and
marketing were beyond the concern of most loggers. The latter
were not bothered by the prospect of drought in Kansas, or chinch
bugs in Dakota Territory, or grasshoppers in Nebraska; their
primary interest appeared to involve little more than the prospects for "a good winter." Furthermore, the capital investment
of the loggers was likely to be small, especially when compared
to that of the sawmillers.

Even if true, such generalizations in no way diminished the
fundamental importance of logging. That Ingram and Weyerhaeuser could justify spending more than a decade following each
other through the Chippewa pineries underlined that fact. But
many lumber producers and most dealers seemed intent on ignoring the entire subject if at all possible. With a uniform lack
of success, Ingram tried to get his Empire business associates to
accompany him occasionally on his logging-camp tours, and not
because he desired companionship. Rather he thought that once
they gained a better understanding of the complete operation
and an appreciation of the costs involved along the way, the
causes for differences over questions of price would be substantially reduced. Although the price of logs accounted for about
two-thirds the price of wholesale lumber, Ingram seemed gen-

[13] See Appendix, Table II, p. 292 below.

erally unable to convince those downriver that their profits and losses depended to a large extent upon the work in the pineries and along the driving streams.[14] Colonel Dulany's response in February, 1888, to such an invitation was what Ingram came to expect: "I would like to go with you to the woods & see some of those big sled loads of Logs But it is too cold for my Rheumatism."[15]

Ingram's impression that logging was a fairly simple business quickly changed with experience in the woods. In the course of the 1880's, he learned something of the less tangible but no less important factors: the isolation of the Wisconsin logging camp, the many dangers inherent in logging activities, just how cold cold could be, the relation of food to the maintenance of morale, the importance of a foreman who could lead and, if necessary, drive his crews, and many other considerations not quite so obvious from behind a desk in Eau Claire.

In many respects, logging camps were paramilitary organizations with an emphasis on discipline and rules aplenty. But rules and discipline could not eliminate the crucial importance of esprit de corps among the members of the crew. Indeed, because they worked and lived in unavoidable intimacy, dissension within a logging crew could have considerable effect on the success and expense of the operation. Thus personality factors assumed greater importance than was the case in many other situations. In this respect the Empire Lumber Company was extremely fortunate to have in its employ John Wight, who served as the traveling woods superintendent for Ingram throughout the 1880's.

Wight roamed the pineries with a critical eye and a pleasant disposition. Unlike Ingram and Weyerhaeuser, he understood logging down to its smallest detail and he was accordingly sympathetic to the problems commonly encountered in the woods. On the other hand, Wight could not be hoodwinked by loggers who often tried to make the most out of apparently unfavorable conditions. Wight's primary responsibility was to insure that the loggers completed their work according to the terms of their contracts. On his inspections from camp to camp, he was not only concerned with the number of logs being banked; he was also

[14] *Idem.*
[15] D. M. Dulany, Sr., to Ingram, February 20, 1888.

concerned with the quality, the grade of logs being banked. Since the interests of log seller and log buyer usually failed to coincide, many camp bosses required encouragement from time to time to increase their efforts or to pay closer attention to one or more particulars. Wight offered this encouragement, hopefully before the arrival of an Ingram or a Weyerhaeuser.

Occasionally John Wight warned a foreman that too much pine was being "left for the worms."[16] More commonly, however, the loggers, who were paid by the thousand feet, tried to include in that amount logs which were convenient to the bank but with little else to recommend them. Wight described one such effort in late January, 1886: "I found those men . . . putting in any thing with sufficient bark to hold together and straight enough to lay still on the landing. . . . If plain talk will do it, I think I made them understand that you did not want every scrub pine on the lands, which they seemed to think you did. I think they will do better."[17]

Almost equaling Wight's concern over the performance of the men in the logging crews was his concern over the performance of the prime movers of the pineries—the teams of draft horses and oxen. This was a most important subject, worthy of a reminder that if the nineteenth century was the age of steam, it was also the age of the horse, and in few places was this more obvious than in the pine forests. Pine logs floated easily, but first they had to be removed from the stump to the stream; and this operation depended almost exclusively on horse and cattle power. When Ingram's son Charles began to assume a larger role in the affairs of Empire, his major responsibility for many years centered on keeping the company well-supplied with good horses. He became the firm's equine expert and he traveled throughout the Middle West checking leads and looking for sound draft animals at fair prices. It was not unusual for horses and oxen to be shipped from points as far distant as Hannibal to be used in the winter logging operations. Yet there always seemed to be a shortage of such animals. One of the reasons for the continuing deficiency was that the horses and cattle were used up quickly by the

---

[16] John Wight to Ingram, March 8, 1886.
[17] John Wight to Ingram, January 23, 1886.

most demanding sort of work under the worst possible conditions. Anyone with a particular fondness for horses could hardly approve of what the logging activities demanded of the beasts. In a sense, horses sent to the pineries were not unlike those unfortunate mounts that picadors ride into the bullring: they were not willingly sacrificed, but their mortality rate was high. Ingram himself was a great fancier of horses, and partly for this reason and partly because they constituted a considerable item of expense he had no patience with those who over-worked the animals or mistreated them in any way. Indeed, few things made him angrier than evidence of carelessness in the handling of horses. But there was simply no way in which logs could be moved to the benefit of the draft animals, and therefore when Wight came across healthy teams at an Empire camp in mid-February, he immediately became suspicious. By that time the work should have taken its toll, with the deteriorating condition of the horses and cattle conversely reflecting the growing size of the rollways along the stream banks. Checking further into a situation called to his attention by fine-looking animals, Wight found that the teams had been hauling loads averaging two thousand feet, which he considered at least five hundred feet less than should have been the case. One teamster in particular, Wight observed, "has got a rong [*sic*] impression of what he is up here for. He thinks it is fat horses, not logs that are wanted and says it will kill the team to do any more. If that be so I think they had better try it [and] if it does kill them it would be better so, than to have them a[n] expense to the company. . . ."[18]

But there were times when even John Wight felt sympathy for oxen and horses, especially when they were forced to fight their way through deep and heavy snows. Such was the situation in January, 1887, when old and doubtless tough cattle steaks became a regular item on camp menus. As team after team literally worked itself to death, Wight wrote the Eau Claire office asking if anyone knew of a "place where you could have ox hides refilled?" He explained that there was nothing wrong with the hides themselves, only that the insides were missing. He had already sent twelve hides down to Eau Claire, he said, and "we

[18] Wight to Empire Lumber Company, February 21, 1884.

have a few more walking around that we may be obliged to send you."[19]

Although Empire's logging superintendent occasionally described such hard times in a decidedly light manner, he tended to reserve that privilege for himself. During the following winter, when deep snows and hard crusts again made hauling agonizingly slow and difficult, Wight did not appreciate Chamberlin's reminder that the Eau Claire office expected a million and a half feet of logs to be banked at an Empire camp near Phipps, Wisconsin. The assistant secretary was reminded in turn, and quickly: "We can easily make figures but it takes lots of hard work to bank a million feet of logs." Wight thought that were Chamberlin afforded an opportunity to see the sorry condition of the cattle for himself, he would be willing to reduce his figures.[20]

Because of changing weather conditions and the corresponding effects upon logging activities, one of the most important determinations a woods foreman had to make was when to work and when to rest. It made no sense to "use up" the cattle and horses during unfavorable periods and thus be in no position to take advantage of improved conditions. That was somewhat the situation Wight encountered in the winter of 1887–1888. In the latter part of March the weather turned cold and road conditions were ideal, by far the best they had been all season. "The greatest trouble," as Wight observed, "is going to be with our oxen, poor brutes, I pity them as they are getting tired and sore. . . ."[21] He doubted that many of the teams would be able to last past the first of April, even under the best of circumstances. But the work was completed, proving the wisdom of the earlier decision not to push the horses and cattle, sacrificing them to little purpose.

Winters are long in northern Wisconsin, and the logging seasons must have seen unendurable at times, especially when work went slowly or not at all. The best loggers tended to be philosophical about many things, but most about the weather. Acceptance of the weather as one variable entirely beyond control came early and contributed towards an ability to maintain good spirits.

[19] Wight to Empire Lumber Company, January 15, 1887.
[20] Wight to Empire Lumber Company, March 14, 1888.
[21] Wight to Empire Lumber Company, March 20, 1888.

John Wight was a great one for shrugging off concerns over too much or too little snow, or whatever other inconvenience nature might impose. The challenge of the task was to succeed in spite of the weather. Thus in early 1888, during a period when blizzard followed blizzard, Wight wrote to a sympathetic Chamberlin that storms had become so commonplace that "when they tell us as they do tonight, 'there is a big circle round the moon,' we say, let her circle."[22]

Although it was not the reason for the trips, Ingram assumed that his frequent appearances in the camps encouraged the logging efforts. He obviously could exert a great deal of influence by the fact of his authority, but he preferred to think that his mere presence in such remote areas promoted increased effort and efficiency by reminding everyone involved of the importance of the work. Furthermore, Ingram thought it must have some beneficial effects on the general morale if the president of Empire occasionally gave the appearance of enduring the same hardships and even eating at the same table with the members of the logging crews. In truth, when Ingram and Weyerhaeuser visited the camps, they did so as army generals, and they no more shared experiences with their loggers than generals can share the danger and privations of their foot soldiers. Both Ingram and Weyerhaeuser demanded special treatment and, of course, usually received it.

There were rare instances, however, when accommodations or attitudes were in some manner deficient, as in the case when Frederick Weyerhaeuser visited an Empire camp in early March, 1890. The camp foreman, Bill Hall, normally slept in a separate little shanty apart from the main building, but when Weyerhaeuser decided to spend the night, Hall moved in with his men, giving his own bed over to the important guest. But it seems that this exchange may not have been graciously made, for Weyerhaeuser subsequently complained that he had overheard Hall remark that "it would be the last time he would give up his bed for somebody else." Among loggers whose enjoyments were few —the right to complain being one of them—Hall's remark seemed harmless enough. Yet Weyerhaeuser chose to hold Ingram per-

[22] Wight to Clarence Chamberlin, January 19, 1888.

sonally responsible for the unfriendly attitudes of the Empire crews, and in this instance, as Ingram informed John Wight, was "much annoyed about it." To provide some salve to Weyerhaeuser's injured feelings, Ingram requested that Wight write a letter for Hall to send to Weyerhaeuser in which an acceptable explanation was devised to give the unfortunate remarks a more agreeable meaning. In these rather silly circumstances, Ingram obviously felt badly, and sincerely so, that one of his own camps had been inhospitable, or at least that Weyerhaeuser had come away with that impression. As he explained to Wight, "I should feel very much hurt, and I cannot believe that Mr. Hall would be so ungentlemanly as to say or do anything to lead Mr. Weyerhaeuser to believe that he would not give him a bed . . . even if he had to go himself into the other camp." He concluded by observing, "I have never found a place in the woods but what they would do that for me and do it ungrudgingly."[23]

In any event, neither Ingram nor Weyerhaeuser was "one of the boys," and it is impossible to imagine that their unannounced inspections of the logging camps occasioned much unbridled joy. Indeed, like military commanders, both could be counted upon to be as stern and unpleasant as the situation seemed to require, and both got results which doubtless had as much to do with rank as with reason.

Proportionally, Weyerhaeuser and "O.H." cut similar figures as they fought through the snows and cold of the Wisconsin winters in order to watch over the work in the pineries. Ingram, at six feet, was the taller by a few inches; weighing close to two hundred pounds, he was also the heavier by several pounds. In general build and the impression of physical strength the two men were much alike. Perhaps Ingram looked a bit more distinguished in part because mustache and chin whiskers, rapidly turning white, received obvious care while Weyerhaeuser's full salt-and-pepper beard was a more ordinary arrangement. But most important, these energetic two were blessed with strong bodies and constitutions during an age when sickness and disease slowed the efforts of so many. Seldom were either Ingram or Weyerhaeuser unable to do as planned for reasons of ill health or injury. In the woods they were a match. Together they directed

[23] Ingram to Wight, March 14, 1890.

the logging operations in the Chippewa valley, and together they learned more and more of the details of this important phase of the lumber business.

In his capacity as director of the executive committee of the Chippewa Logging Company, Ingram may never have spent a long enough unbroken stretch in the woods to gain a full appreciation of the importance of the logging camp cook. Aside from their obvious need for nourishment, an especially large requirement for the hard-working loggers, the men had very little to look forward to other than meals. If the food was inadequate or poorly prepared, the crew was unhappy; and if the crew was unhappy, the logs arrived slowly at the rollways. These workers enjoyed few rights, but some unwritten law seemed to guarantee them the privilege of good and sufficient food.

Good cooks were not an abundant item. All too frequently changes in kitchen personnel had to be made in the course of the season, occasionally disrupting the progress of work in the woods. In the 1886–1887 winter, for example, N. B. Noble, logging superintendent for the newly-organized Rice Lake Lumber Company, reported to Ingram that one of the crews was "very much exasperated and excited and did not work Monday," because the "cook at Bear Creek became unbearable."[24]

Although meals were very important, there is little evidence that the loggers were especially hard to please. After a time, they even came to accept the use of oleomargarine in the Wisconsin pineries. But there were always a few who passed themselves off as cooks and who soon proved otherwise. A foreman for the Daniel Shaw Lumber Company reported the presence of just such a cook in his camp, tactfully inquiring if Mr. Shaw thought it paid "to hire this kind of man to come up here to learn to cook." In this instance, there was no complaint about the personality of the cook, "but as for him being a cook, he fails very much." Shaw's foreman continued, offering specific illustration:[25]

> I will mension a few things. In making gravy for potatos he would season it with all kinds of esence. No man could eat it and he tried to make mince pies. He made them all out of

[24] N. B. Noble to Ingram, February 2, 1887.
[25] Samuel Hubble to Daniel Shaw, December 29, 1873, in the Shaw Papers, Eau Claire Public Library manuscripts collection.

meat. They made good meat pies if every one liked them dry. This is a specimen of his fixing up every thing.

Largely because of the difficulty in procuring enough male cooks with reasonable skills, females were occasionally hired for the work, almost always the wife of the foreman or one of the crew members. As the years went by, it became more and more common for wives and entire families to join their husbands deep in the pineries, although the privilege was generally limited to the management echelons of the business. John Wight took a very dim view of this trend. He maintained that neither women nor liquor advanced the cause of logging. Wight told Charlie Ingram of an incident involving Bill Hall, the same foreman who offended Weyerhaeuser. Wight described a vigorous discussion he had recently held with Hall on the subject of women in the woods, in which he had told Hall: "A man must have little respect for a wife to bring her into such a place as this." Only then was he informed of the impending arrival of Mrs. Hall. But John Wight did not feel particularly apologetic for his faux pas. As he explained to the young Ingram, the more he thought about the Hall incident the more he became convinced that he should have been even stronger in his denunciation of the practice. With disgust, he described the Empire Company tote roads as resembling "a moving indian village covered with women & kids."[26]

Indeed, during the 1888–1889 season, the situation became so serious that it could no longer be ignored. Wight, with the support of Charlie Ingram, decided to eliminate the camp followers at the Lost Lake Camp and subsequently use it as an example for all Empire camps. Accordingly, the tote man received strict orders to take no unauthorized persons, specifically women and children, back into the camp. These instructions were largely ignored, however, and in the process the problem was seriously complicated; not only were women in the camp, but they were there in direct contradiction of orders. Clearly Wight could not allow this insubordination to pass without response and remain as logging superintendent. As he wrote on February 1, 1889, "I shall be ashamed to go there after this and have them fellows feel they had gained a victory over me." He closed his

---

[26] Wight to C. H. Ingram, January 31, 1889.

letter to Charlie asking for assistance: "I am ashamed, disgusted and badly beat and shall await your answer as to how yourself and father feel would be the right course to take with them." Response from the Eau Claire office was immediate. Within the week, the foreman at Lost Lake was fired and it was once again apparent to all that in Ingram's absence, John Wight spoke with authority in the camps of the Empire Lumber Company.[27]

The difficulties involving females and families in the logging camps seem to have been largely of a temporary nature. The same was not true of problems relating to alcohol. By official policy, no lumber companies permitted liquor in the logging camps, but in the case of Empire, the rules were enforced with great dedication. Ingram himself assumed the major responsibilities of watch-dog in this matter, readily accepting such added burdens as part of his Christian duty to his men. In the view of Empire's president, abstinence possessed an even higher virtue than efficiency in the logging operation, though in truth the two were inseparable.

From all accounts, there were relatively few violations of the liquor ban in the camps themselves, but control of neighboring establishments proved to be a more difficult task. Still the pressures which Ingram could bring to bear on the proprietors of tiny saloons scattered through the woods were considerable. Generally, when he learned of any of his men being served liquor, he would threaten the provider with a boycott. In the isolation of the pineries, customers were already scarce and the possibility of losing all of the Empire trade gave most proprietors cause for reflection. In some ways, however, the problem was without solution. If an owner of an inn or saloon knuckled under to Ingram and stopped selling alcohol, he might continue to do a little official business with the Empire Lumber Company but often not enough to compensate for his losses. This was the dilemma facing W. E. Cornick, who operated the Spider Lake Summer Resort near Hayward, Wisconsin. Cornick had reluctantly followed Ingram's advice and had not sold any liquor for two years. The result of the change in policy, as he complained, was that "[I] have lost all the lumber mens traid . . . That class of People are bound to go where there is Liquor." It was not strictly a question of

---

[27] Wight to C. H. Ingram, February 1 and February 10, 1889.

wanting to sell the intoxicating beverage, Cornick admitting that his wife agreed with Ingram on the subject, yet it seemed possible that "Sircumstances will drive me to it."[28]

Although Ingram and Wight could not avoid spending much of their time dealing with problems associated with the Empire loggers, the pine logs remained their crucial concern and the reason they spent their winters in the Wisconsin woods. In the case of Ingram, his interest in pine logs ranged far beyond the camps of the Empire Company, involving all of the Chippewa Logging Company operations, other camps of member firms, and those of any independent loggers who had contracted with the Pool.

If there were some who knew more than Ingram about the details of logging, very few knew more about sawlogs. Judgment in this matter was the special province of the sawmiller. He knew, from long years of experience, how the logs "opened up" or what quality of lumber could be expected from a particular kind of log. There were many, among them Frederick Weyerhaeuser, who regarded the president of Empire as the best judge of sawlogs in the Chippewa valley. But few of Ingram's Empire associates were convinced that those skills were so unique as to justify a sixty-year-old man spending so much time and energy traveling from camp to camp and exposing himself to extreme discomfort and, indeed, danger.

Colonel Dulany, in particular, took Ingram to task on many occasions over the question of these logging-camp inspection tours. In a sense, old Dan resented the fact that the president had become so vital to the Empire Lumber Company. Too often

---

[28] W. E. Cornick to Ingram, November 15, 1896. Personally Ingram was not the absolute abstainer that most presumed. He thoroughly enjoyed a glass of Madeira and a good cigar while relaxing with male friends and associates, discussing business and politics. Although such occasions were rather special, it was certain that Cornelia Ingram was never able to accept indulgences of that sort, even if they brought her hard-working husband pleasure. Perhaps because of Cornelia's disapproval, Ingram endeavored to keep his social drinks both occasional and private. After all, he was a leader of the church and an example for the community.

Grandson Charles Ingram recalls an incident when his grandparents and he were on their way up to the summer cottage at Long Lake, near Rice Lake, Wisconsin. In the course of this trip, one of the Ingram suitcases was accidentally dropped on a railway station platform and a bottle inside was broken. The resulting odor left no doubt that Ingram had lost a special beverage, but he quickly remarked for all to hear, "Why it seems that we have broken our bottle of Listerine."

it seemed that without Ingram there would be no Empire. From the vantage of Hannibal, this consideration made the trips into the pineries more than a mere matter of personal foolishness; they amounted to an inability on the part of Ingram to recognize the nature of his primary obligations to family and friends, and, of greatest importance, his responsibilities to Empire. Over and over again, the Dulanys attempted to set him back on the true course.

In February, 1890, while the Colonel was enjoying a rest at the health resort of Hot Springs, Arkansas, he received a letter from Ingram recommending that Dan push the worries of business entirely from his mind, at least for the duration of his stay at the spa. Colonel Dulany found such counsel from one like Empire's president amusing, and he responded that it seemed most unlikely, were their situations reversed, that Ingram would be able to follow his own advice. Once started, Dulany had no trouble easing into the real subject of his concern, observing that[29]

> . . . custom has made you such a slave to business, that you think business, eat, sleep, and dream business, and then almost break your neck, and some of these times will break it, if you are not more particular, running through the woods to the different Logg camps to look them over to see just the kind of·loggs they are cutting, and the length of each Logg, and whether you think that logg will make strips, Boards, Dim., or Shingles. now there is no use in you exposing yourself and taking the risk of your life and health in running over these camps to see what kind of loggs they are cutting. wait until you get those loggs into the Mill boom, and then decide what you will cut them in to . . . you will doubtless say that somebody must look after the Logging & Camps. We are willing to admit that, but it does not necessarily follow that you are the only one that can do that, and if I thought you was the only one that could perform that duty, we would want to commence looking towards the end of our business because the time is comeing, and I fear not far hence, if you don't stop and take care of yourself you will not be able to rip around over the woods looking after the logging.

Ingram tended to resent the raising of the subject of his winter inspection tours, and for a number of reasons. To begin with, they had to be among the most unpleasant and wearing part of his responsibilities. Also, despite many invitations, none of his

---

[29] D. M. Dulany, Sr., to Ingram, February 24, 1890.

associates, including the Hannibal Dulanys, had been able or willing to share the experience of a winter trip through the pineries. As a result, in matters relating to logging, they remained largely uninformed. The primary reason for Ingram's special sensitivity to the subject of camp inspections, however, involved Weyerhaeuser. Colonel Dulany naturally liked to think that he and his brother were reasonably well-informed about the entire Empire operation, but in truth there was much that they did not know or did not understand, particularly about logging activities and arrangements. Old Dan had long been in the habit of offering advice as to how best to get along with Weyerhaeuser, but the problems of that relationship were far more complicated that the Dulanys could imagine. Not only were they naive when it came to the subject of Weyerhaeuser and the Pool arrangements, the Dulanys really could not be of much assistance anyway. Consequently, Ingram had been reluctant to provide them, especially the Colonel, with additional ammunition and cause for concern to no apparent purpose. He therefore tried to explain his own activities and responsibilities in logging matters without reference to Weyerhaeuser. These incomplete explanations never quite rang true. After receiving the Colonel's letter from Hot Springs, Ingram did decide to discuss the actual issues, the real reasons he continued to drag himself through the Wisconsin woods.

In no way deprecating his own skill in judging logs, the president of Empire admitted to the Colonel that he could doubtless teach another, "especially a young man," to do nearly as well. But that was not the problem which most concerned him. While another might indeed develop the ability to judge the different lots of logs in much the manner of Ingram, it seemed quite unlikely that another would have much influence with Weyerhaeuser "when the time came for making prices on those logs." It was this consideration which more or less dictated the nature of Ingram's activities during the logging seasons beginning in 1881–1882. As long as he maintained such a large investment in the Chippewa Logging Company, Ingram could not allow himself to "lay back and submit to what I and others interested with me would be obliged to if I did not take some of these trips," or so he explained to partner Dulany.

Weyerhaeuser was the one to whom submission would be complete were Ingram to give up his log-inspection activities. It was not simply a question of whether or not Weyerhaeuser intended to manipulate matters to his own advantage, particularly as concerned the determination of log prices. As a matter of fact, Ingram agreed that "in the main" Weyerhaeuser seemed "inclined to look at such matters in a reasonable way. . . ." But inclinations were not enough. Ingram maintained that for all of Weyerhaeuser's good intentions, it was simply not possible for him "to divest himself of the interest he has in those logs," adding:

> He is interested in more logs on the Chippewa directly and indirectly than any man I know of . . . I am disposed to give him credit for putting in as a rule good logs and having them graded very closely. While I am willing to do that, I know he is very tenacious about having all they are worth, and I think sometimes he gets more than they are worth. If Mr. Weyerhaeuser would delegate his work to somebody who has no interest in the logs you may be very sure that I should delegate these trips of inspection to some younger person than myself.

With the passing seasons, Ingram became increasingly selective concerning which logging camps he visited. The Chippewa Logging Company employed inspectors such as Ed Douglas and Henry Morford, whose reports sufficed in nearly all cases where the price of logs had already been established by the terms of a contract. Ingram restricted his own visits to those camps where the prices had not been fixed, a determination which would subsequently be made by the executive committee of the Chippewa Logging Company. Sometimes these camps were strictly an outside operation then doing business with the Pool, but more commonly the camps to which Ingram gave careful attention were operated by the member firms, the stockholders of the Chippewa Logging Company. As he explained to Colonel Dulany, of the logs in which the Pool was interested with price still to be determined, "quite a large percent of them are being put in by Weyerhaeuser & Denkmann & Rutledge, Weyerhaeuser, Denkman & Laird, Norton, Weyerhaeuser & Rutledge, and Weyerhaeuser, Laird, Norton & McCord, and Weyerhaeuser & Sawyer County Bank, and Weyerhaeuser, Brewster & Miles." This was incomplete, Ingram observed; the entire list would be considerably longer, but he had only wanted to provide Dan with a general idea of the scope of Weyerhaeuser's logging interests.

One might presume that the increasing extent of these logging interests in the Chippewa valley and elsewhere would tend to lessen Weyerhaeuser's personal management of the details of operation to a significant degree. That, however, seems not to have been the case. Despite the hard work of travel, accommodations less than satisfactory along the way, and the difficulties imposed by unfavorable weather, Ingram readily admitted that his German friend kept up with things "pretty sharpe." It would have made life a good deal easier for Empire's president, and for many others, had Weyerhaeuser been less vigorous and therefore less able to traverse the pineries with such thoroughness. But, as Ingram observed, Weyerhaeuser was "a very hardy, tough man, hence it does not seem to worry him very much to do it."

Easy or difficult, pleasant or unpleasant, the fact that Weyerhaeuser gave so much of his time and energy over to matters related to logging provided an indication of the relative importance of that phase of the lumbering operation. This was exactly what Ingram had long been trying to get his own business associates to understand. In another attempt to gain this end, Ingram reported to Colonel Dulany the details of an encounter on the train with Weyerhaeuser in early January, 1890, in which the latter had complained that the return on his sawmilling investment for the 1889 season had not been more than 3 per cent. In the course of their conversation, he showed Ingram a balance sheet for Weyerhaeuser & Denkmann indicating a profit for the year of $192,000. "That one," Weyerhaeuser admitted, "looked a little better." Ingram pointed out for the benefit of Dulany that the reason for such a considerable profit was that the logging operations in which the Rock Island partners held an interest had been closed out into the Weyerhaeuser & Denkmann account, "and as I have said frequently before, I think they have made more money out of logging than they have out of their mills, especially [in] late years."[30]

Thus, despite dangers, discomforts, and pneumonia "following along at your heels ready to take hold of you at any time,"[31] as long as Weyerhaeuser insisted on traveling from camp to camp

---

[30] Ingram to D. M. Dulany, Sr., March 3, 1890.
[31] D. M. Dulany, Sr., to Ingram, March 10, 1890.

and taking a direct hand in setting the prices on logs, Ingram could do no less. Both had too much of their own invested to allow the other complete freedom in such matters. Accordingly, when "Mr. Wherehauser [sic]" and Captain Henry visited the Empire camps, making "free use of their pencil as usual," carefully inspecting the rollways and charging that the logs were badly cut, graded, and scaled, John Wight was not overly concerned. As he remarked after one such visit: "I am satisfied . . . and I think Mr. Ingram will be."[32]

[32] Wight to Empire Lumber Company, February 16, 1887.

CHAPTER IX

Empire Expands

FOR ALL OF Ingram's many involvements with sawlogs and with Frederick Weyerhaeuser, and the adjustments he made and responsibilities he assumed within the Chippewa Logging Company, there were also responsibilities to meet and adjustments to be made within the Empire organization during the decade of the 1880's. One of the more obvious requirements for change had to do with the ability of Empire to continue to supply their old wholesale customers. Ingram was not anxious to break off these relations, but it was becoming increasingly apparent that the Eau Claire mills of Empire would be hard-pressed to meet the needs of their own wholesale yards at Winona and Hannibal.

During the winter of 1880–1881, Ingram had sought to appraise these long-time friends and customers of the recent changes in organization, asking that they place their orders early should they be wanting any lumber from Empire during the coming season. Dealers like Henry Dortland of Glen Haven, Wisconsin, received letters in January of 1881 from the Eau Claire office explaining: "We will not have but a few customers outside of our own yards now and would like you to be one of them if you would like. . . ." Under such arrangements the lumber ordered would be priced at the market rates at time of delivery, and any cash advances would be accepted at 7 per cent interest."[1] Even though the resulting promises made to outside dealers were few, they were honored only with difficulty. Despite Ingram's reluctance, in future seasons Empire lumber went strictly to Empire dealers.

The sawing season of 1881, the first for the new Empire Lum-

[1] Ingram to Henry Dortland, January 25, 1881.

ber Company, began well, and by July 1, after about two months of steady work, nearly 65,000 logs, more than 13 million board feet, had passed through the saws of its four Eau Claire mills. The Big and Little Mills, at the old site between the river and Half Moon Lake, had sawed 5,115,510 and 1,922,440 feet respectively; and the Eddy Mill and Dells Mill, both located upstream above the Dells Improvement Company dam, had handled logs amounting to 3,751,400 and 2,645,730 feet respectively. Within the new organization, the sawmilling responsibilities were soon divided so that the Eddy Mill sawed almost exclusively for the Winona yard and the Big Mill for Hannibal. Ingram assumed that, aside from dimension lumber and shingles, these two mills could produce lumber approximating the amounts required by the two wholesale operations. Also, by assigning a separate mill to each yard, Horton and the Dulanys could deal with a single superintendent, allowing for greater efficiency in sawing to the particular needs of the Winona and Hannibal markets.

After such a good start in 1881, however, plans and production suffered a setback in July as a result of the familiar problem of low water in the Chippewa along with a relatively new kind of trouble in Eau Claire. The millworkers struck for a ten-hour day, apparently to the great surprise of Ingram and his fellow millmen who, of course, saw no legitimate cause for complaint. The so-called "sawdust war" was short-lived, ending when Governor William Smith called out the militia in defense of law and order. As a result of a compromise on an eleven-hour day, violence was averted, and before the end of the month Chamberlin was able to inform the Hannibal Dulanys that the Empire mills were back in full operation. Most agreed that they would probably be able to saw as much lumber in eleven hours as had previously been accomplished in twelve.[2]

Business was good at the Hannibal end of the line during the summer of 1881. In the heat and humidity of August, the crews pulled and piled Empire lumber at an average rate of twenty-five cribs, or 250,000 feet a day. By August 12, eighty-five carloads of lumber had been shipped and orders for a hundred more remained on the books. The Dulanys had just sold a bill for

[2] See the *Eau Claire Free Press*, July 21, July 18, and August 4, 1881; Vernon H. Jensen, *Lumber and Labor* (New York: Ferrar & Rinehart, 1945), 59.

nearly three-quarters of a million feet to the Union Pacific Railroad and, as the Colonel reported, "we are driving along as best we can under all the disadvantages of hot weather, dust, teams & men." In part because of the extreme heat, they had difficulty in hiring teams that summer and were forced to buy a number of horses in order to keep abreast the raft arrivals, moving the lumber from the levees to the yard. Accordingly, the Colonel advised Ingram that before he or anyone else in Eau Claire purchased teams with a view towards winter logging, they should first check with Hannibal since, in all probability, "we can supply you."[3]

But Colonel Dulany's concerns with heat and horses were passing things. A far more persistent source for worry, and a cause for prolonged discussion and disagreement in the seasons ahead, were the policy decisions made in the Eau Claire office on such matters as lumber prices and the even more basic question of Hannibal's role within the Empire organization. To an extent, the effectiveness of the senior Dulany's recommendations relating to management decisions was blunted by his own lack of discrimination. The Colonel tended to complain about everything, forcing Ingram to try to sift the important from the unimportant. On most occasions, however, Ingram doubtless knew when Dulany was genuinely upset about something; and Dan was genuinely upset in the late summer of 1881, that initial season of Empire's operation.

Complaints from the wholesale yard men could usually be assigned to one of four categories: the price at which lumber was billed; the assortment, or the varying percentages of boards, strips, and dimension lumber; the grade or quality; and delays in the arrival of rafts. Occasionally the reason for complaint was neither caused nor correctable by humans. For example, when low water interfered with rafting operations, even Colonel Dulany saw little purpose in blaming those at the Eau Claire end of the line. As he wrote to Chamberlin in the summer of 1886, when the stage of the Chippewa prevented rafts from reaching Reads Landing, they were all patiently waiting in Hannibal since "we are not disposed to fight nature." In this instance, even old Dan had to admit that he saw no way for Chamberlin to raise the level of

---

[3] D. M. Dulany, Sr., to Ingram, August 12, 1881.

water in the Chippewa, "unless you get all your people, when they first get up of a morning, to go up on to the Dells Pond and. . . ."[4]

More often than not, however, the cause for complaint did involve human error, or at least what were viewed as errors from the vantage of Hannibal, Missouri. Such was the case in 1881, when the chief complaint centered on differing views over the price at which the lumber was being billed to the Hannibal yard. In August, the Colonel told of "killing ourselves to put in thirty million ft.," which he thought they would sell without difficulty. But, he added, had they been buying the lumber outright on the market at the prices billed from Empire's Eau Claire office, he doubted that they would have wanted more than 20 million feet. Thus was introduced what would prove to be a chronic point of contention within the Empire Lumber Company organization.

The fundamental question was a simple one: At which end of the line were the profits to be credited? To some, such a difference might seem moot, but not to the Dulanys. Indeed, the Colonel acknowledged that Ingram could ask why the price of billing mattered much since "the result will be the same so far as making money is concerned when we wind up our years business." In reply, Dulany argued that each point ought to be given an opportunity to make its fair share of the profits; that there was no accurate method for testing efficiency of operation unless each department of the company maintained the notion of competition against its rival organizations. In other words, the profit motive had to be allowed to function at all points and not just in the Eau Claire office. As Colonel Dulany attempted to explain to Ingram,[5]

> . . . we think our folks could work with a better grace this very hot dry weather, if they knew our lumber in the yard was not costing us any more than our neighbors were costing them for the same quality and that they had a chance to make the same profit that our neighbors have, and if we are on even footing with them we can hold our own.

Although he did not disagree with the Colonel's views on this

---

[4] D. M. Dulany, Sr., to Chamberlin, August 7, 1886.
[5] D. M. Dulany, Sr., to Empire Lumber Company, August 12, 1881.

particular subject, Ingram tended to disparage the seriousness of differences which developed between the separate departments within the Empire Company. In any event it certainly was not his intention to bill the Empire wholesale yards at anything but fair and competitive rates for Empire lumber. In addition, he clearly recalled, even if the Dulanys had forgotten, that Ingram, Kennedy & Company and Dulany & McVeigh had not always agreed on what constituted fair and competitive prices in the past. Consequently Ingram was not too surprised or disappointed that these differences continued even after they had become business partners in Empire. He knew that old habits were hard to break.

Nevertheless, the areas of disagreement that became so quickly evident between the Eau Claire office of Empire and their men in the field could not easily be dismissed. They were more fundamental and complex than a mere concern over the prices at which lumber was billed, although this subject, by virtue of its concreteness, often became the focal point of contention.

As is so often the case with growing institutions, the officers of Empire were experiencing what seemed to be a built-in malfunction, having less to do with questions of policy than with matters of size, distance, organization, and personnel. Basically, the problem was one of maintaining a spirit of unity within a large and very diverse organization. The fact that the Dulanys knew almost nothing about the logging and sawmilling operations contributed significantly to the development of antagonisms between staff and line. Even the exchange of personnel, such as the appointment of Dan Dulany, Jr., to the Eau Claire office staff in 1882, where he assisted Clarence Chamberlin, seemed to have small effect on these internal differences. It was in this area of organizational relationships that Orrin Ingram gradually became aware of the new nature of his own responsibilities. Clearly he was a participant in the management revolution—perhaps none too willing, but a participant nonetheless. His involvement in the production and sale of lumber had become increasingly indirect, and he was now far more concerned with those who produced and those who sold.

At times it must have seemed to the president of Empire that his primary function was to convince each member of the organization that everyone else was doing his job; or, as the Dulanys preferred to phrase it, that the Eau Claire officials were making sacrifices equal to those being made in Hannibal. Colonel Dulany, gloomy so much of the time, wrote often of their depressed condition, the result of unremitting demands made upon the Hannibal yard for money needed by the Eau Claire office, usually for purposes he considered ill-advised. During such periods, Dan's talent for humor could not veil the seriousness of his complaint, the essence of which was that the Eau Claire personnel worked less, made fewer sacrifices, and enjoyed greater benefits than their colleagues in Hannibal. At one point he suggested that it would prove helpful if the boys in the home office could be made to realize that writing checks was a far more pleasant activity than maintaining the balance. To provide just such a lesson, the Colonel thought it might be a good idea if they could all arrange to change places for a time. His band from Hannibal would go up to Eau Claire and begin spending money while the Eau Claire office crew went down to Hannibal to "enjoy the fun of raising money to meet the company's obligations." The old gentleman added: "You bet we would make it lively for you and let you realize for a little while what we have to go through all the time."[6]

To emphasize their hard-pressed circumstances downriver, in the fall of 1884 Dulany complained that because of the financial situation of the company they could no longer even provide themselves with life's barest necessities. "I am without a horse," the Colonel grumbled, "and have to foot it to the river and all over Town, because we haven't the money to spare out of the business to buy one."[7] Dan suspected—and probably correctly— that no one footed it very far in West Eau Claire. In fact, the matched teams which pulled his carriages were always a special matter of pride to Orrin Ingram. Of more importance, however, the Dulanys were simply experiencing some of the more unappealing aspects of Ingram's policy of insisting that all available funds be employed to advantage, even if it meant that Empire's vice-president was occasionally forced "to foot it."

These lamentations were repeated regularly and with only slight variation when slow sales, pine-land purchase, or large

[6] D. M. Dulany, Sr., to Empire Lumber Company, September 27, 1884.
[7] *Idem.*

payments due created temporary shortages in the Hannibal cash drawer. In September, 1889, for example, Dulany told Ingram of their impoverished condition, reduced to "two meals per day and no ice in the water. . . ." A week later the Colonel thought Ingram might be interested in the reasons he and the boys had been unable to attend the Hannibal fair, always a big event in those parts. In the first place, it had been unseasonably cold and they had no overcoats to wear. But even had the weather been warmer, they could not have afforded the admission charge because they had "got right down to hard Pan, and if any of you have any old clothes, Hats or Boots to spare, send them to us."[8]

As the second largest stockholder and vice-president of Empire, in addition to being the manager of Empire's largest wholesale yard, it was natural for Colonel Dulany to expect that his recommendations and requests would receive appropriate consideration. Thus, when it seemed impossible for Eau Claire to provide lumber in satisfactory condition, or in the kinds and amounts requested by the Hannibal yard, he tended to become emotional. Early in the 1883 season, disgusted with the condition of the bundles of shingles arriving on the Empire rafts, old Dan admonished young Dan and Chamberlin to spend less time "playing with your pretty wives" and more time down on the loading docks.[9] In the case of the Colonel's namesake, reproval of this sort must have brought a smile, for he had recently wed, against the strongest wishes of the Hannibal homefolks, one of Eau Claire's most charming and successful courtesans.

During the previous summer, the Eau Claire office had learned that the Hannibal yard was entirely out of 4 x 4's and had been forced to purchase from neighboring firms all of that size needed to fill orders. The boys in the home office were subsequently called upon to answer the question, *"what do you think of a yard that's interested in several Mills and have to buy every 4 x 4 that they use?"*[10] Indeed the problem of keeping inventories current was one of continuing concern, and the changes suggested by Colonel Dulany brought considerable improvement to an area

[8] D. M. Dulany, Sr., to Ingram, September 18, 1889; Dulany to Empire Lumber Company, September 25, 1889.

[9] D. M. Dulany, Sr., to Empire Lumber Company, June 21, 1883.

[10] D. M. Dulany, Sr., to Empire Lumber Company, July 11, 1882.

that had been neglected by the Eau Claire office. But the one modern means of communication which could have eliminated most of the inventory-production difficulties—the telephone—was never used at the Hannibal end. In the early 1890's, when Ingram suggested that they consider installing a phone at the yard office, the Dulanys decided, not untypically, that they were too near closing up the business to warrant such an expense.

In his capacity as wholesale dealer, however, the Colonel's principal cause for complaint sooner or later returned to the subject of the price at which Empire lumber was billed to the yards. A typical instance occurred in the summer of 1883 when, without discussion or even any prior notice, the Eau Claire office advanced the figures on August rafts. Dan Dulany had at least two good reasons for reacting to this sudden change in prices. The first was a simple difference of opinion as to whether such an increase was justified; the second was the arbitrary nature of the decision for an advance. He regarded the second consideration as the more serious.

Why, he inquired of Ingram, should he in Hannibal and Charley Horton in Winona feel unhappy in their relations with Empire Lumber Company home office? The reason could not involve their share of the ownership, for they both held "a respectable portion of the stock," and it might reasonably be assumed that they would share somewhat equally in making important decisions as well as in profits and losses. But that assumption, they had all come to realize, was not entirely correct. Furthermore, referring specifically to the recent advance in billing prices, the Colonel again insisted that it was not enough to say, "what difference does it make . . . it all goes into the Empire Lumber Company." If that remained the line of reasoning in Eau Claire—that it made no difference which end made the profits—then the decision to raise the price of the August rafts would seem to make even less sense.

Colonel Dulany attempted to point out that most of these differences were avoidable. The real source of irritation was not the change itself but the method of change. Dan explained that if these decisions had been made in a spirit of co-operation, "then we would not consider ourselves as mere hewers of wood, and drawers of water, but [as] partners in the Empire Lumber Company." He then summarily closed the discussion by informing the boys in the home office that the raft had been charged up "at old prices."[11]

Often, in the wake of serious disagreements of this sort, Clarence Chamberlin devoted more than normal time to Hannibal correspondence. With great patience he would attempt to salve the wounds, carefully explaining just what considerations had prompted the most recent changes in price, assuring the Dulanys that times were generally hard, and that no more was being asked of Hannibal than of other points along the line. The Colonel much appreciated Chamberlin's thoughtfulness, and through the years he grew fond of the "old brick," his favorite nickname for the hard-working assistant secretary of Empire. But even Chamberlin's tact could not prevent periodic outbursts. Thus in mid-summer, 1885, old Dan erupted anew, again over the question of an advance on Empire lumber. Not only did he "most emphatically . . . protest," he also informed Eau Claire that it was impossible to permit the lumber to be entered at those prices, explaining, "we don't believe in making fictitious or false entries on our Books. . . ."[12]

A few seasons later Ingram attempted to eliminate the causes of such grievances by delaying, until the close of the year, the decision on wholesale prices. Then, with all of the figures available, Eau Claire and Hannibal representatives could meet together and determine a fair average price on lumber. But even this solution failed to satisfy the Colonel who, by late August, 1889, considered his earlier acceptance of that new arrangement to have been "a big mistake." Ever since the agreement to delay price determination, he complained to Ingram, the Eau Claire mills had sent down to Hannibal rafts which contained "Clear and Culls, sap, rot, worm Eaten, Norway, Hemlock, White Pine, Guts & Feathers all in the same crib."[13]

All complaints from Hannibal could not be taken seriously. As Ingram quickly learned, the Colonel was only happy when he had something to complain about and he never seemed to lack raw material. For the most part, Ingram silently and patiently accepted this wailing. In truth, the pessimism of the

---

[11] D. M. Dulany, Sr., to Empire Lumber Company, September 1, 1883.
[12] D. M. Dulany, Sr., to Empire Lumber Company, July 10, 1885.
[13] D. M. Dulany, Sr., to Ingram, August 31, 1889.

Dulanys was often considered less a cause for alarm than a source of amusement. No matter how bad conditions might become, one could be certain they would never approach the disastrous levels predicted by Colonel Dulany and his brother William. A letter written by Horton to Chamberlin in 1886 could have been written with equal application during the majority of Empire's seasons: "As usual Hannibal has done much better than we expected—doubt that they will have to sleep under the Bridge and drink branch water for a while yet."[14]

But if complaints from Hannibal were seldom discouraged, they also had relatively little effect. Ingram may not have been by nature autocratic, although with the passing years he would acquire a manner and a carriage which left no doubt as to his authority; yet it was true that on most business subjects—not including negotiations with labor committees—Ingram was willing to tolerate all manner of discussion and listen to suggestions from a variety of sources. After completion of the preliminaries, however, he was generally accustomed to having his own way in business decisions involving those firms in which he was the majority stockholder. Not all of his partners and business associates cared to dispute matters of policy with Ingram. Charley Horton, for one, seldom discussed his differences directly with the president of Empire. Rather he tended to send any complaints he might have to Colonel Dulany, fully confident that they would be passed along to Eau Claire. Dulany, quite obviously, required no intermediary. His complaints, as noted, were direct and frequent. There were also instances when Major Day at Dubuque objected loudly and lengthily about an error in judgment or a difference in opinion.

Although the privilege of complaining was generally observed and maintained, in the majority of situations the act of complaining provided sufficient satisfaction, or at least the matter ended there. This is not to say that the meetings of the directors of Empire were meaningless affairs, or that Ingram would insist on a particular course of action when all others were in opposition. But there was no doubt whose opinion weighed most heavily, and when Ingram decided that the time had come for a de-

---

[14] Horton to Chamberlin, March 15, 1886.

cision, to all intents and purposes discussion ended. Neither Dulany nor Day, and certainly not Horton, was interested in drawing the line on Ingram. It was always his approval that was sought and his disapproval that was feared. Few details seemed too small to escape his interest, and when Ingram wrote a letter, ostensibly "for purposes of information," there was no question but that the subject of the correspondence received more than usual attention in the days to follow.

W. H. Day was Ingram's business partner longer than anyone else, but he never quite became an equal. Even after assuming the presidency of the Standard Lumber Company, the Major rarely did anything of importance without first seeking Ingram's advice and approval. He even hesitated to take a vacation until he was sure that "O.H." would have no strong objection. Thus, when planning a trip east with his wife to visit her "homestead" in the summer of 1884, Day wrote to Eau Claire: "I have thought that without very serious business sacrifices I might be spared for three weeks, but I would like an expression from you."[15] Ingram knew that Major Day was not simply being polite.

On the few occasions when the downriver partners erred in prolonging discussion past the point of decision, Ingram did not hesitate to remind them of their relative positions. In the late summer of 1884, for example, after Ingram had determined that the Empire mills should end their sawing season earlier than usual, he did not appreciate an additional prod from Hannibal questioning the propriety of that decision. He reacted with unusual irritation, telling his Hannibal associates to tend to their own business responsibilities and leave the decisions involving production to those who knew something about it. Old Dan accepted this dressing-down in good spirit, reminding Ingram that he had only suggested a reconsideration, and that he fully appreciated the limitation of his own abilities and experience. But the Colonel's retreat was orderly, and he observed: "we think we know the demands and requirements of this end of the line," just as Ingram knew the demands and requirements at the other end. In conclusion Dulany expressed his complete confidence that Ingram could reach the correct decision re-

[15] Day to Ingram, July 20, 1884.

garding shutting down the mills without assistance from Hannibal since, as had clearly been stated, "we do not know anything about it. . . ."[16]

Colonel Dulany did know something about selling lumber, however, and during the 1880's Hannibal became the largest consumer of Empire pine lumber. Although demand and prices varied slightly from season to season, in general more and more lumber was produced and marketed. In 1881, the first full year of the new Empire organization, unfavorable stages for rafting and other factors combined to frustrate plans to yard upwards of 30 million feet at Hannibal. Only slightly more than 24 million feet of Empire lumber reached the yard, supplemented by some 2 million feet purchased from other sources. Of the total, almost 23 million feet had been sold by the end of the year. Profits at Hannibal for 1881 amounted to $127,000, which Dan Dulany announced was rather below his expectations, a fact not apparent from his predictions earlier in the season. It was nevertheless no small achievement for, as the Colonel explained to Ingram, they had worked "under more difficulties than ever before, paid higher for labor, and got less work, more bother from high water, deep mud and everything else, than in any former years. . . ." All of this had been bad enough, but Dan warned everyone to prepare for much worse times ahead, because, "We see the spirit of Communism is looming up in all directions."[17] Between the communists and the coming panic— and there was always a panic coming—it was understandable that the Colonel accepted success uneasily.

In the course of working through the Empire books prior to the start of the 1882 rafting season, and no doubt with Colonel Dulany and the controversy of the preceding August over lumber prices foremost in mind, Chamberlin came up with an estimate of production and delivery costs per thousand feet of lumber. These figures he made available to all of the Empire stockholders. Allowing for a probable tendency to inflate a bit here and there, Chamberlin's computations remain of interest. The cost of stumpage, or the pine trees without the land, he estimated at $3.00 per thousand feet; and the expenses of logging, including

---

[16] D. M. Dulany, Sr., to Ingram, September 3, 1884.
[17] D. M. Dulany, Sr., to Ingram, March 14, 1882.

the hauling from stump to stream bank, $4.00. Scaling, driving, sorting, and booming costs added another $1.47, all of which amounted to $8.47 per thousand feet for the sawlogs as received in the mill ponds at Eau Claire. These figures were somewhat complicated, however, by the fact that not all of the logs cut in the woods reached the mills. According to the statistics available to Chamberlin, an average of only about 65 per cent of the sawlogs were received the same season they had been cut. About 20 per cent would reach the mill booms during the second season and a few more arrived the third year after cutting. It was further estimated that 10 per cent of logs cut and banked never reached their downstream destination. Figuring the value of the sawlog at an even $8.00 per thousand feet and annual interest at 8 per cent, Chamberlin calculated the costs of delay in delivery at nineteen cents per thousand, with an additional eighty cents per thousand for those 10 per cent of logs forever lost.

As to the sawmilling expenses, these Chamberlin estimated to average $3.00 for each thousand feet of lumber. Construction of the cribs and running the Chippewa rafts from Eau Claire to Reads Landing added another seventy-five cents; and Mississippi running from Reads to Hannibal, including breakage, $1.75. Thus the total cost per thousand feet of lumber pulled at the Hannibal levee, as estimated in the Eau Claire office, amounted to $14.96. Whether the Colonel accepted the "old brick's" figures as gospel is, of course, doubtful. In any event, the Dulanys could not fail to appreciate that there was considerably more to the lumber business than wholesaling in Hannibal, Missouri.

But for all that had gone before, the final and most important object for the entire Empire organization was customers, prices, and sales; and the manner in which the 1882 season opened made more than a few wonder if the return would be worth all of their effort and expense. Dulany described himself as "the highest Bull in the place," although in its attempt to hold the line on prices, the Hannibal yard lost sales to such an extent that by quitting time on April 19, no orders remained on the books. The Dulany brothers had to go back three years to find a day so "blue." In view of these circumstances, the Colonel expressed some second thoughts about a project approved at the recent annual meeting of Empire—the construction of a sawmill at Winona. Dulany now indicated that they were all "a little nervous about it, owing to the high price and scarcity of stumpage," although he was quick to add that Ingram would, of course, know more about such matters than they.[18]

Concern over the condition of the early season markets and unfavorable driving conditions was temporarily overshadowed by a late April fire which swept through the West Eau Claire milling district. Empire losses turned out to be comparatively small, amounting to some $36,000, of which $20,000 was covered by insurance. The loss actually involved little more than the old Ingram, Kennedy & Mason lumber store, a point of contention dating back to the days of Alexander Dole. Now it was Colonel Dulany who expressed small regret over the path of the fire, explaining, "We were all a little anxious to close out the Store, and quit that part of the business. . . ."[19]

If the store made little sense to Dulany in the 1880's, in the earlier years it had served a purpose of real importance. In those days, the credit of the young company was not always above question and the goods required in their operation not always readily available. The decision to establish a store had resulted primarily from the consideration of logistical questions, and in these areas the store provided its lumbermen-owners with greater flexibility and greater security. Cash or not, credit or not, vital goods could be obtained; and it did not take the lumberman long to discover that syrup for the table was as necessary as handles for the axes.

The profits from the store had never been large, and of themselves hardly warranted its continued operation. Yet there were occasions when the store provided another small but important source of ready cash; and as Ingram had long since learned, even a little cash was critical at times, for a little cash could often buy considerable credit. But these were lumbermen, not grocers, and it was certainly not Ingram's intent to become a grocer on even a part-time basis. The store was important, as the bank was important, because both contributed to the success of the lumbering operations. If such involvements were "side shows," as Dulany commonly contended, the indirect benefits had nevertheless been significant.

---

[18] D. M. Dulany, Sr., to Ingram, April 20, 1882.
[19] D. M. Dulany, Sr., to Ingram, April 27, 1882.

By the time of the fire in 1882, however, the company store had indeed outlived its original purpose if not its usefulness. By then there were numerous stores and the procurement of goods had become a relatively easy task. The reputation of the Empire Lumber Company made all storekeepers anxious suppliers, on terms of cash or credit. But it became apparent that it was much easier for Colonel Dulany in distant Hannibal to advise that they cease operating "that silly store" than it was for Ingram to follow the advice. Considerations other than those of strictly a business nature apparently complicated Ingram's attitude towards the store, by then an institution in West Eau Claire, making the decision to close it down more difficult. The April fire quickly and efficiently did away with the need for a decision.

Of greater importance, although similar to an extent, was the disagreement over the continued involvement of Empire with the Ingram, Kennedy & Gill Company yards at Wabasha and neighboring towns in southeastern Minnesota. Immediately following the merger into Empire, the Dulanys had expressed considerable doubt about maintaining business relations with both the Wabasha and Dubuque operations, although the profits subsequently announced by Major Day soon limited their objections to Wabasha. In the Colonel's opinion, Ingram, Kennedy & Gill would never amount to very much, "and if the lumber you left there had been sent to Winona, and us, we would have made more out of it." This letter, written before the store was destroyed in the April fire, counseled that the Empire management ought to rid itself of the Wabasha affiliation, "together with the Store . . . and anything else that don't legitimately belong to our branch of business and let us run that for all that is in it, and our ability will let us."[20]

As had been the case with the lumber store, however, any decision affecting Wabasha could not be made on the sole consideration of business factors. There were people involved, people that Ingram cared about and felt responsible for—most importantly, the widow and family of William V. Gill. Ingram attempted to explain these matters to late-comer Dulany, stating, "it would not be treating those that are interested with us just right for us to sell out." The Colonel replied that he too under-

[20] D. M. Dulany, Sr., to Ingram, January 11, 1882.

stood and supported the Golden Rule, but this did not prevent him from opposing the continuance of unprofitable business associations.[21]

At the center of this difference of opinion was the lumber which was permitted to go to Wabasha. Since there seldom seemed enough to go around, Dulany contended that Wabasha was an expensive diversion; that much more could be realized from the Empire lumber were it allowed to come down to Hannibal. He argued that Wabasha was simply too close to the mills to serve in any efficient manner as an independent distributing point. Furthermore, the Wabasha market was obviously restricted by Winona and Minneapolis dealers on either side, both points enjoying the advantages of larger and busier rail connections. In a sense, Wabasha seemed a duplication of effort on two counts. Not only could Charley Horton's Winona yard supply many of the points served by Wabasha, but dry lumber could be shipped by rail from Eau Claire at no great expense. After considering the costs of pulling, washing, hauling, piling, and selling the river lumber, the advantages of the short water route between Eau Claire and Wabasha appeared unimportant.

Yet again it was easier to talk about these "facts" than it was to sell out the Empire interest in Wabasha and the little cluster of tributary retail yards. Actually, despite all the attention given the subject, the involvement at Wabasha was not large. By the middle 1880's, the Ingram, Kennedy & Gill partnership was shared by the Empire Lumber Company, Mrs. Gill, George DeLong and Charles Chamberlain; Empire controlled six-fifteenths of the business, Mrs. Gill owned four-fifteenths, and DeLong and Chamberlain one-sixth each. Empire's portion of the 1886 dividends amounted to $1800, a piddling amount to be sure, but at least Wabasha was not a losing proposition.[22]

In any event, Ingram subbornly or loyally resisted all efforts to end the Empire association with Ingram, Kennedy & Gill until 1887. Early in that year, DeLong and Chamberlain persuaded Mrs. Gill to sell out her interest to them.[23] It seems apparent that Ingram did not entirely approve of the action, though he

[21] D. M. Dulany, Sr., to Ingram, January 24, 1882.
[22] Ingram, Kennedy & Gill balance sheet, December 31, 1886.
[23] George DeLong to Clarence Chamberlin, January 13 and February 4, 1887.

made no effort to interfere. Through the years, in his relations with the Gill family, he had been a good friend and a conscientious guardian. With Mrs. Gill's decision to sell out her interest in the company, Ingram felt relieved of the necessity for further personal involvement. Although he would maintain a financial interest in the Wabasha Lumber Company, successors to Ingram, Kennedy & Gill, it was an interest of a much simpler sort than had previously been the case.

But in 1882, Wabasha was only one of many problems confronting the management of Empire and, most particularly, vice-president Dulany. As noted, old Dan was a great one for scanning about and finding dark clouds on the business horizon. A crash, a panic, a major depression was always just around the corner and the constant admonition from Hannibal was "we had better be in shape. . . ." Had the grand strategy been left to him, the Empire Lumber Company policy would have concentrated on getting through the hard times in the best possible manner with only incidental attention to the intervening periods of large demand and good prices.

Ingram's ideas were the exact reverse. Predicting panics was like predicting rain: sooner or later the prophet would be proved correct. But so far as the lumberman was concerned, the period taken as a whole was one of prosperity, not of depression. Accordingly, the more logical policy was one that reflected the general trend rather than one designed to meet the exigencies accompanying a panic whose date of arrival, although inevitable, was unknown and unknowable. The overriding consideration, as Ingram viewed these matters, was to take the utmost advantage from the many seasons of prosperity and be satisfied with enduring, one way or another, the so-called hard times, always keeping in mind that depressions were the exception. Indeed, it was an expanding economy in a rapidly expanding nation, and there continued to be a tremendous demand for lumber.

These basic differences in business philosophy were at the crux of many of the disagreements between the Dulanys and Ingram, just as had been the case earlier in disputes involving Ingram and Dole and, to some extent, Ingram and Kennedy. During all but the best of seasons, every investment seemed like a speculation to Colonel Dulany, and he maintained that the privilege of speculation was a luxury only accorded the sound firm. Ingram simply could not make the old gentleman understand that if "sound" had to mean being out of debt, then the Dulanys would need either to change their attitudes or change their business.

On the one hand, Dulany argued that it made little sense to consider the possibility of expanding sawmilling capacity when they already lacked sufficient pine land or stumpage to supply the mills then in operation.[24] But when Ingram made the necessary purchases of pine land, he could expect loud and long complaining from the one who advised Chamberlin, not to rest "until you get the last copper paid off."[25] Faced with such a limited opportunity to please his Missouri partner, it is small wonder that Ingram proceeded to do largely as he himself thought best. Furthermore, the Colonel was notably inconsistent in the positions he assumed with so much spirit. When sales were slow, he knew that further investments ought to be avoided, but when demand was large he was less certain. Such a variable guide was not of much use to the president of Empire.

But if Colonel Dulany was not always pleased with the decisions made in Eau Claire, he found their subordination to Frederick Weyerhaeuser even more distressing. From the beginning of his involvement in the Chippewa Logging Company, Weyerhaeuser had insisted on a free hand in the negotiations for pine land and property. He had wasted no time in considering purchases of large extent and expense. For example, on April 12, 1882, he proposed to purchase nearly 200,000 acres of Cornell University land containing an estimated 844,423,000 feet of pine at a total cost of $2,632,030, with an option for an additional 25,000 acres containing another 84,551,000 feet. As would be the case in all such purchases, payment was to be "guaranteed by all the firms composing the Chippewa Logging Co. in proportion to the stock held by each firm in said company at the date of final contract."[26] The possibility of the powerful Weyerhaeuser increasing Empire's indebtedness on a mere whim hung heavily with the Colonel. He passed by no opportunity to remind Ingram that the Chippewa Logging Company had not been

[24] D. M. Dulany, Sr., to Ingram, August 12, 1882.
[25] D. M. Dulany, Sr., to Chamberlin, December 22, 1882.
[26] Copy of proposed contract, Weyerhaeuser to Sage and Boardman, April 12, 1882.

established to serve the interests of Eau Claire sawmillers. This was one subject on which Ingram needed little in the way of reminders

Demand in Hannibal was slow to improve that season of 1882, and in early June Dulany recommended that both Ingram and "Mr. Weyerhaeser" take a trip down the Mississippi, "and look things over . . . and if you do think it will take all the wind out of you, so far as buying Pine Lands, buildings and buying Mills are concerned." While admitting the possibility that conditions might improve, to Dan it seemed more likely that their bad situation would worsen. According to the information received in Hannibal, only "in Dakota and Manitoba" was the demand for lumber reasonably good, and as Dulany observed, "they can't take all the lumber that is made, and my prediction now is that we have not near touched bottom yet." In conclusion, he stated his opinion that it would be the height of foolishness to proceed with the plans for building the Winona mill, at least "until we see which way the cat jumps." If the weather continued favorable, the Colonel predicted August sales from the Hannibal yard would total about 3 million feet. The weather was favorable and shipments exceeded 3 million feet, making the sales for August, 1882, the most "we have ever made in one month."[27]

Furthermore, for all of the complaining, prices were reasonably good in Hannibal. In the middle of August, for example, William Irvine finally managed to dispose of an immense quantity of Chippewa Lumber & Boom Company lumber, some 13 million feet, for prices ranging from $11 per thousand feet for culls to $17 for boards and strips. Although old Dan frowned, shook his head, and announced, "it is very evident they can't make a cent upon it," Irvine and others knew better. It was very clear "which way the cat had jumped," but the Colonel restated the "unanimous feeling" from Hannibal that any plans for building the Winona mill should be postponed for at least another year. By that time, he told Ingram, they ought to have some idea "how things pan out."[28]

Ingram must have wondered just how much evidence was required before Dan Dulany became convinced that times were

[27] D. M. Dulany, Sr., to Ingram, June 3, August 12, and September 4, 1882.
[28] D. M. Dulany, Sr., to Ingram, August 19, 1882.

good. In the meantime, Charley Horton reported from Winona that they were shipping an average of 100,000 feet per day out of the yard.[29] Major Day at Dubuque seemed less concerned with lumber sales than with keeping his sawmill supplied with logs. During the 1882 season, the Standard Lumber Company sawed considerably more than 20 million feet, and in October, the Major went up the Black River where he purchased some 30 million feet of small pine at a price of about $4 per thousand feet.[30]

In the fall of 1882 it was apparent that the Empire mills could not keep up with the demand, and in early November, Dulany and Cruikshank negotiated a joint arrangement with Olds & Lord of Afton, Minnesota, to saw some lumber for them during the 1883 season, Empire's share amounting to between 3½ and 4 million feet. Even fair prices, large demand, and insufficient supply brought no change in the Dulanys' attitude; they continued to oppose any expansion of sawmilling capacity, specifically by the construction of a new Empire mill at Winona and the establishment of an entirely separate operation on the Red Cedar River, the Rice Lake Lumber Company.

The interest in Rice Lake and the Red Cedar resulted more from a lack of alternatives than any other factor. In the early 1880's, business was too good and Ingram was too young to think in terms of a gradual cessation of lumbering activities. But in order to continue production at the established level, much less to provide for any increase, more timber had to be acquired—and timber was rapidly becoming a scarce item in the valley of the Chippewa. This was the situation in September of 1882 when Ingram and his old friend William Carson boarded an eastbound train in Eau Claire with two objects in mind: first, they were to accompany thirteen young ladies, including Mirriam

---

[29] Horton to Ingram, October 24, 1882.

[30] W. H. Day to Ingram, October 12 and 21, 1882. This purchase serves to underline the advantages of doing business in the manner of Weyerhaueser. The pine stumpage that Major Day purchased along the Black River would cut a few sawlogs of fair size—that is, 4½ to 5 per thousand feet—but the majority was much smaller timber, up to twelve logs per thousand. In any event, from a consideration of quality, the price was quite high. In Weyerhaeuser's large transaction with Sage, chairman of the Cornell University land committee, finally approved on August 1, 1882, more than 100,000 acres containing nearly 600 million feet of pine was purchased for $1,841,746. This figured out to $3 per thousand feet for the pine and fifty cents an acre for the land. Hidy, *et al., Timber and Men,* 84. See also Gates, "Weyerhaeuser and the Chippewa Logging Company," 61–62.

and Fannie Ingram, to Wellesley College; and second, they hoped
to arrange for the purchase of some convenient pine land.

For once, even Colonel Dulany gave his blessing to the purchase effort. Writing from Hannibal on the last day of August,
Dan was almost effusive in praise and good wishes, noting that
Ingram had managed the log supply portion of the business during the current season in a "first rate" manner. In addition, the
Colonel admitted to being "glad" that Ingram was setting off for
the purpose of acquiring some pine along the Eau Claire River,
expressing the hope that the deal might be completed within a
few weeks.[31]

As a member of the Weyerhaeuser Pool, the Empire Lumber
Company had as a primary objective the maintenance of a position of self-sufficiency—that is, Empire could guard against serious
risk of loss by putting in approximately the amount of sawlogs
that it took out. Such parity would afford protection regardless
of the prices Weyerhaeuser might arbitrarily establish. Thus reasoned Dulany; and although Ingram recognized the desirability
of timber purchase, Chippewa Logging Company or no, there
seemed little reason to argue the matter. Hopefully some pine
could be reasonably gotten, and hopefully all of the Empire stockholders, Colonel Dulany included, would be able to justify the
investment.

Initially Ingram thought that after he and Carson had safely
delivered the girls to college, they would go on up to Maine where
they would negotiate for some timber which ex-governor Abner
Coburn owned along the Eau Claire River. But en route during
the Chicago stopover, they learned that Coburn was asking
$650,000 and that he was not inclined to permit any opportunity
for examination before purchase. Furthermore, reliable word
had it that the old Maine politician had already promised the
tract to Thorp's Eau Claire Lumber Company. So, as the train
headed eastward out of Chicago, "O.H." and his friend may have
looked like concerned chaperones but were really worried lumbermen, thinking more about pine land than their noisy young
charges.

Alternatives to the Coburn land did not come easily to mind.
As the discussion continued and various opportunities were noted,

---

[31] D. M. Dulany, Sr., to Ingram, August 31, 1882.

both Ingram and Carson probably began to think like younger men again. Why restrict their considerations to the Chippewa valley? Certainly they could command as much money and more knowledge than the great majority of their competitors, present and to be. In this spirit, they turned their attention to a sizable tract on the Red Cedar River which was owned by a New York speculator, William Griffin. For the purpose of sounding Griffin out, Carson left the train at Troy, New York, while Ingram and the girls continued on to Boston. After a good deal of bargaining, Ingram and Carson succeeded in obtaining an option to buy from Griffin. Subsequently that timber provided the basis for the establishment of the Rice Lake Lumber Company.[32]

As expected, this development met with little enthusiasm among the Hannibal folks. While they had acknowledged the desirability of the procurement of additional pine land in the Chippewa valley, such purchases in other valleys seemed to serve no purpose other than to increase the debt at the very time when concerns ought to center on winding down the operation. In addition, an increase in sawmilling capacities as an accompaniment to an increase in timber holdings failed to alleviate the situation in any relative sense. The more saws you had the more sawlogs you needed. Colonel Dulany had long since tired of running on that sort of treadmill and could not understand why he was forever being forced to explain the obvious to Ingram. But Ingram appreciated one important factor that apparently had escaped his Missouri partner—this particular treadmill made money, and the demand for lumber was unlikely to diminish simply because the supply of Chippewa pine did so. The Colonel, however, continued to see the hole in the doughnut, thinking more about debts than profits. In truth, outstanding bills piled high on the desks of every operator, successful or otherwise, but outstanding bills were something that could keep old Dan awake night after night, even though he arose most mornings a wealthier man.

The declared reasons for opposition to the new large investment plans at Winona and Rice Lake differed somewhat. At Rice Lake, Ingram envisioned the establishment of a large milling facility which would ship its lumber to market entirely by rail.

[32] For the details of this transaction, reference may be made to the remarks which Ingram delivered at a testimonial banquet given in his honor at Rice Lake. See the *Rice Lake Chronotype*, December, 1909.

There were, however, some complications to this scheme. A large portion of the timber Ingram hoped to purchase for the Rice Lake operation had been recently sold by the Wisconsin Railroad Farm Mortgage Land Company to the Chicago, St. Paul, Minneapolis & Omaha Railroad Company, whose more common designation, the "Omaha," belied its pine-land origins. Ingram was apparently not much concerned about the details of that land trade, although it had been brought into question by many.[33] He was concerned, however, by the insistence on the part of Omaha officials that as a condition to the sale of land to the Rice Lake Company, their road would be guaranteed the privilege of hauling all the Rice Lake lumber to market.

But simple concern in the Eau Claire office about the details of such arrangements not uncommonly developed into extreme anxiety in Hannibal. When the Colonel learned of the Omaha conditions for sale, he could only reply: "Good Lord deliver us from any entangling alliances with R. Roads. . . ."[34] But the more Ingram thought the matter through, the less he worried about the possible adverse effects of the sale conditions. It seemed unlikely that the provision stipulating exclusive shipping privileges to the Omaha Road would be enforced even if it could be, especially if the Rice Lake Company made every reasonable effort to honor the spirit of the clause.

In February, 1883, Ingram met Colonel Dulany in Chicago to discuss in detail the plans for Rice Lake and any differences they might have regarding the pine land purchases then under consideration. Dulany returned to Hannibal armed with maps, plat descriptions, estimates, locations of lakes, streams, and railroads, and proceeded to present "the matter fairly in all its phases" to his business associates. The vote following his presentation of the case was unanimous in its opposition to the pine-land purchase. In reporting their decision on the matter, Colonel Dulany briefly summarized the reasons for rejecting Ingram's recommendations. In the first place, there was considerable doubt that sufficient investigation of the land had as yet taken place: The available estimates as to the amount of standing pine were those

[33] See, for example, Richard N. Current, *Pine Logs and Politics: A Life of Philetus Sawyer, 1816–1900* (Madison: State Historical Society of Wisconsin, 1950), 133–143.

[34] D. M. Dulany, Sr., to Ingram, January 20, 1883.

of the seller and not their own, and in Hannibal they were unwilling "to make such a leap in the dark" without knowing in how much and what kind of timber they might land. By far the most important objection continued to be the conditions of the sale as stipulated by the railroad. The Colonel maintained that should they insist on their privileges of hauling Rice Lake lumber, the Omaha officials "could control and manipulate our purchase just as it might suit their wishes." For anyone knowingly to place himself in such a circumstance would be sheer lunacy. As old Dan viewed the matter, "the writer would not give his great Grandfather such power over him, much less a Rail Road Corporation."[35]

Despite continuing opposition from downriver, Ingram proceeded with plans and purchases in the Rice Lake area; and in early 1884 he officially organized the Rice Lake Lumber Company, assuming the office of president himself. Long-time friend and fellow Democrat William Carson was the vice-president; Clarence Chamberlin, the secretary; W. K. Coffin, who had been Ingram's clerk in the bank, became the treasurer; and a new associate, N. B. Noble, was the assistant secretary. Noble proceeded to Rice Lake where he assumed command, supervising the construction of logging roads and dams to be used in the driving operations. Construction of the sawmill would be delayed until the fall and winter of 1886–1887, but there was much to be done before production began.

Although the Rice Lake Company did occasionally experience problems of the sort feared by the Dulanys concerning the Omaha and its hauling privilege, these were not sufficient to discourage the rapid growth of the operation. By the 1891 season, its two mills sawed upwards of 30 million feet of lumber and shipped its product to nearly every state west of the Alleghenies.[36] In the early spring of 1892, Colonel Dulany seemed to have come full circle in acknowledging that Rice Lake had proved itself a success. He sent a letter off to the Rice Lake manager inquiring if he might be interested in borrowing some money from the Hannibal bank in which Dulany held a major interest, adding, ". . . would like to let you have all you want."[37]

[35] D. M. Dulany, Sr., to Ingram, February 22, 1883.
[36] A. L. Ulrich to Ingram, October 3, 1881.
[37] D. M. Dulany, Sr., to Rice Lake Lumber Company, March 5, 1892.

In retrospect, reviewing the development of such successful operations as the Rice Lake Lumber Company, the Dulanys appear to have suffered from a considerable lack of foresight. Their extreme conservatism made them unduly conscious of pitfalls and dangers, thereby making large decisions especially difficult. In a sense, they seemed unwilling to press their luck. In 1883, for example, in the midst of Ingram's Rice Lake considerations, Colonel Dulany tried to make him understand the Hannibal view of things, observing that the Empire Lumber Company was already "a pretty large thing," and assuming the continuation of reasonably intelligent management and economy, "in the course of ten years think we will all be in comfortable circumstances." Such expressions of timidity were not entirely new to the president of Empire. Indeed, former associates had left him for those same reasons, at a time when he was far less successful.

But Ingram himself remained the principal cause for concern and for reluctance on the part of the Dulanys to expand their personal investment or to increase the indebtedness of Empire. Dan assured Ingram that if they had any guarantee that Empire's president would remain healthy and active for "12 or 15 years we would not fear spreading out so much but the only guarantee we can have of this is we don't believe you will die or get killed if you can help it. . . ."[38] That was not quite enough assurance to cause the old gentleman to rest easily.

This basic insecurity naturally colored any consideration of expansion, and the Dulanys already possessed sufficient reasons to question the direction of Ingram's leadership. The construction of a mill at Winona provided yet another opportunity for disagreement between the opposite ends of the Empire organization. Proposed and approved shortly after the 1881 merger, the plans for Winona had been tabled in the course of the 1882 season, in large part because of objections from Hannibal. By 1885 Ingram seemed to have altered the original proposal, doubtless in the hope of gaining the support of the Dulanys. The substance of the new proposal involved removing the Little Mill machinery from Eau Claire to Winona, simply constructing a new building around it. Thus it appeared that there would be no change in the total sawing capacity of Empire, only a change

[38] D. M. Dulany, Sr., to Ingram, January 20, 1883.

in the sawing location of one of the Empire mills. The Colonel understood, however, that this was little more than a device to obtain his vote. Ingram had no intention of closing down the Little Mill completely, although in the future production activities there would be limited to sawing twenty-four-foot and longer logs into piece stuff. But this limitation would not eliminate the continued need for the carriages, edgers, and trimmers along with the rotary saw, leaving Dulany confused over just what machinery Ingram planned on removing to Winona. The Colonel also realized that once a new mill was in operation, the door would be opened to unending requests for improvement and enlargement. No sawmiller that he had known was ever long satisfied with his plant.

Dulany made no effort to forward a timely response to Eau Claire offering his own views on the subject of the Little Mill removal to Winona. Ingram waited impatiently for an answer. He knew that the delay was not the result of any uncertainty or indecision on Dulany's part. Far more likely the Colonel had formed an answer immediately after reading over the new proposal, making his procrastination appear, at best, inconsiderate.

Finally, prodded by an angry letter from Ingram, Dulany sat down on the last day of September to put his thoughts to paper. He did so with no hint of apology for his tardiness, chiding Empire's president, "We did not notice anything that required an immediate reply from us." The suspicious and sensitive old man then observed that since everyone in the Eau Claire office assumed the Hannibal crew "knew but very little, or nothing about building mills . . . that we would not venture an opinion in regard to these matters." Typically, however, the Colonel could not resist venturing an opinion, and Ingram read on to learn that the Dulanys were totally unimpressed by his amended proposal for a Winona sawmill:[39]

> . . . In view of the shortage of stumpage which is growing rapidly less every year, and more difficult to procure & control, and we thought we had more mill capacity than the future outlook for a supply of logs would justify—under this state of facts, would advise as much economy as possible in removing or improving the Mills we now have, so as to let the Mills & Stumpage wear out about the same time.

[39] D. M. Dulany, Sr., to Ingram, September 30, 1885.

Viewed from such a vantage, Empire had already begun its decline. Ingram realized that a mill at Winona was important for the very reasons offered by Colonel Dulany in opposition. Without the new mill, the Empire Lumber Company would indeed have to begin the process of closing down as white pine became increasingly scarce and expensive in the valley of the Chippewa. A milling facility at Winona, however, would permit the sawing not only of logs from Chippewa pineries but logs from the upper reaches of the St. Croix valley. The logging frontier had only recently moved with serious purpose into the stands of pine in extreme northwestern Wisconsin and northeastern Minneosta at the western end of Lake Superior, and much of that region was drained by the St. Croix and its many tributaries. There pine land could still be secured, and it was this pine that Ingram planned to saw at Winona.

The idea of a railroad yard with an adjacent mill such as he planned for Winona and Rice Lake also must have had some appeal. Ingram left no evidence that he worried much over the question of the so-called rail differential, a freight rate imposed in the early 1880's which was decidedly unfair to the Eau Claire producers who shipped by rail directly to western retail markets. But the fact that the differential seems to have been no great issue with him, and that Empire lumber from Eau Claire mills would always leave in rafts, does not mean that Ingram was blind to the advantages of rail shipment. They were all too evident, and were probably becoming more so with the passing seasons and the gradual realization that more of their markets were urban and their customers were more particular.

As long as the relative costs remained competitive, Ingram certainly had no objection to allowing the logs to move on water and the lumber on rails, arriving at market clean, white, bright, unwarped, and unstained. As for the lumber produced in the Eau Claire mills of Empire, however, he remained unconvinced that the benefits of rail transportation were sufficient to warrant any change in his own well-established patterns of marketing. Also, the Hannibal customers seemed none too demanding. After all, yellow pine was fast proving to be a saleable item in those parts. Consequently the lumber pulled by Empire crews at the Hannibal levees continued to be stained, streaked, and silt-covered; but it was also cheap, and with some scrubbing and planing looked little the worse for the long trip.

In his plans for the Winona mill, as had been true with the Rice Lake operation, time again proved Ingram correct. The Eddy Mill at Eau Claire was closed following the 1886 sawing season and superintendent Will Tearse proceeded to Winona where he assumed charge of the construction of the new facility, using as much of the Eddy materials and machinery as could be profitably employed. The first whistle blew on the morning of May 20, 1887, and in less than two weeks the able Tearse informed Ingram that he was ready "to hurry matters up a little before long now."[40]

It was clear that for some time the Eau Claire officers had felt that the Winona branch of the Empire organization suffered from poor management. Horton was amiable enough, but he had grown old too early. Tearse, therefore, had a more complex assignment than merely watching over activities inside the mill, although in its early days that was a large responsibility in itself. But Ingram also hoped that Tearse would be able to provide some life and vigor to the Winona operation in addition to his sawmilling expertise. The superintendent appreciated the delicacy of his new situation, trying to please Ingram without displeasing Horton, and he promised to "do what I can to make the investment here as good as I know how to & hope we shall catch up within speaking distance in the race with our very able & competent managers at Hannibal & Dubuque."[41]

"Catching up" with Hannibal and Dubuque was no mean task. As an indication of what Tearse and Winona faced in the matter of internal competition, out of the 36,428,037 feet of lumber sold by Empire of Eau Claire during the 1886 season at an average price of $11.30 per thousand, Winona yarded only about one-third of the total, or 12,214,796 feet. The Hannibal share amounted to nearly 23 million feet, and Wabasha received the balance, or slightly more than 1,300,000 feet. Sales at Hannibal totaled $379,732.20 for the season. The average cost of lumber in the water at that point was computed to be $11.29; piled in the Han-

[40] Tearse to Ingram, June 1, 1887.
[41] Tearse to Empire Lumber Company, January 13, 1887.

nibal yard, the cost averaged $12.09; and the average sale price per thousand feet was $15.20. Thus, on lumber alone, the Dulanys counted a surplus of income over cost of nearly $70,000. Major Day enjoyed comparable success at Dubuque. The Standard Lumber Company mill sawed slightly in excess of 22 million feet that season, manufacturing an additional 1¼ million shingles. The 1886 balance sheet listed a surplus of $61,211.11 which, on a total capitalization of $300,000, amounted to a net gain of more than 20 per cent. Returns like that made both Ingram and Day pleased that they had persevered through less pleasant and less profitable periods.

Despite the knowledge and business abilities in obvious operation at other points along the line, the availability of pine in areas accessible to Winona assured eventual success for the new sawmilling facility there. Although the logs were smaller and less perfect than those Tearse had become accustomed to back at Eau Claire, the Winona mill sawed and sold lumber in increasing amounts, exceeding 20 million feet by the 1890 season. More important than its early success, however, was the consideration which had prompted Ingram's initial interest in the construction of a sawmill at that point. Winona was the only Empire mill still sawing lumber at the turn of the century, its pine resources in Douglas County in extreme northwestern Wisconsin providing logs through the season of 1908. The Empire was also the last of the Winona sawmills to cease operations, Laird Norton having closed in 1905 and the Winona Lumber Company mill the following year.[42]

But this end of operations was local and had been long foreseen. Like so many other frontiers, that of the lumbermen moved again, the only difference being the distance of the move and the age of the Lake States operators, many of whom were too old, or too weary, or perhaps even too wealthy to begin anew.

[42] See the *Winona Republican-Herald,* November 15, 1906, and November 14, 1907.

CHAPTER X

The End of Empire

IN THE FOUR DECADES following his arrival in the Chippewa valley, Orrin Ingram had come a very long way, measured both in terms of dollars and from a consideration of methods of operation and organization. He might have gone even further had he and the circumstances been somewhat different. The Weyerhaeuser interests were then in the process of another merger and a great move to the forested slopes of the Pacific Northwest. Ingram could have accepted a position of prominence within the Weyerhaeuser Timber Company's table of organization, but he chose not to. However much he had been able to adjust as a younger man in the acceptance of new roles and responsibilities, at the end of his career he was unwilling to make another move and begin all over again in an unfamiliar area and working for a company not his own. He was old, and his ties in Eau Claire were strong. Furthermore, Ingram had never quite been able to become the corporation man, at least not with any conviction and enthusiasm. He was a transition-type of entrepreneur who had succeeded, with some difficulty, in moving about half-way from the free and independent days of partnership.

His "one foot in and one foot out" compromise arrangement between the Chippewa Logging Company and the Empire Lumber Company was an interesting one, but clearly Ingram had accepted it largely through lack of alternatives. A move west with Weyerhaeuser at the turn of the century, however, was another matter. Nevertheless, the decision to remain in the Chippewa valley and outside the active Weyerhaeuser management was not altogether an easy one. In effect it meant that Ingram's involvement in lumbering had come to an end.

The season of 1892 proved to be the last of the big years for the Empire Lumber Company, and for many other business organizations as well. The panic so long predicted by Colonel Dulany was truly close at hand. Although lumber continued to be sawed and sold in the seasons to follow, for the duration of the nineteenth century and well into the twentieth the work somehow seemed much harder as margins became much smaller. Whether the decrease in standing pine or in eager customers was the more to blame for the end of prosperous times in the lumber markets is difficult to say. Certainly the sudden decline in demand initially brought things up short in 1893, and the decline in stumpage made any subsequent recovery short-lived, at least in the upper Mississippi valley.

Also difficult to explain is why the Panic of 1893 affected Midwestern lumber markets so much more severely than had the earlier depressions. The difference may well have had something to do with the changing nature of those markets. Perhaps far more than Ingram and other producers had realized, their 1890 customers enjoyed greater flexibility in the matter of purchasing lumber. In 1873, for example, the farming frontier had been expanding at a rapid rate, but by 1893 this expansion had slowed. Clearly the picket fence around a city house was not the same critical item as the corn crib on a pioneer farm.

Logging conditions had been generally favorable throughout the Lake States pineries during the Winter of 1891–1892. But what had begun as one of the most promising business years literally became mired in the mud of a wet, cold, and depressing spring. The rains came and continued with such persistence that by late May no lumber had been pulled from the Mississippi at Hannibal, and few fields of corn had been planted on Missouri farms. The cold rains of Missouri and Iowa occasionally changed into snow and sleet over Minnesota and Wisconsin. In a rare occurrence, the Winona mill closed down on the afternoon of May 19 because of the inclement conditions, and the following morning Tearse awoke to see the Wisconsin bluffs across the river again white with snow.[1] In Rice Lake the weather was even more severe, a blizzard forcing temporary suspension of sawmilling and yarding activities.

The weather continued depressing for so long a period that

---

[1] Will Tearse to Ingram, May 20, 1892.

spirits as well as business suffered. Colonel Dulany seemed on the verge of announcing the end of the world. On May 21 he notified Ingram that, "taking all things together the outlook has never been so bad since we commenced business as it is today."[2] But Ingram had heard similar lamentations from that direction on so many previous occasions that he gave them little notice. As the months passed, conditions did improve, and at the end of the season Dan admitted that Hannibal had not done so badly after all. Less lumber, by some 5 million feet, had been sold in 1892 as compared to 1891, but the prices were better, and "Our collections have been very good."[3]

In fact, problems had become so scarce at the end of the 1892 season that the primary concern among the Hannibal Dulanys seemed to be one of finding a use for all of their surplus capital. Simply counting it afforded little real pleasure, even for one as conservative as the Colonel. He described their affluent circumstances, observing, "We have now paid off all our Bills Payable and will have to find some way of using the money that will be coming in." He hastened to add, however, that a solution to their over-abundance might already be on the way. The Dulanys were expecting the arrival of Charles Horton that same day, and Colonel Dulany, still sensitive to some recent investments in pine land and logging railroads for Winona, predicted that Charley "probably can suggest a plan."[4]

Ingram was in a situation similar to that of the Dulanys, experiencing for the first time in his life some uncertainties about where to invest his surplus funds. The reason for this unusual position was clear enough. The old opportunities for investment no longer existed. The Empire Lumber Company made a small purchase of Cornell University land as late as December, 1892, but this was among the last such trades available within the Chippewa valley.[5] Although some additional pine land and stumpage purchases were made in Douglas County, these were charged to the separate Winona account.

Empire ceased to expand when the opportunities for expan-

---

[2] D. M. Dulany, Sr., to Ingram, May 21, 1892.
[3] D. M. Dulany, Sr., to Chamberlin, November 28, 1892.
[4] *Idem.*
[5] E. L. Williams, treasurer of Cornell University, to S. Robertson, December 19, 1882.

sion ceased to exist; specifically, when pine land was no longer available in the Chippewa and neighboring valleys. With all the bills paid and little use for surplus funds, the management of Empire had no choice but to increase the size of its dividends, thus giving back to the individual stockholders the responsibilities for investment decisions.

As a result, Ingram may have had more money than he could wisely manage alone, at least in areas outside those associated with lumbering. His personal interests became extreme in their diversity, ranging from coal mines in the Canadian Rockies to orange groves in Florida. The mining interest in particular— soon to involve lead, gold, copper, and anything else of value, in addition to the anthracite coal—seemed somehow out of character for the old lumberman. For all of his optimism, Ingram could hardly be described as a gambling type or the sort who would be attracted to the get-rich schemes of the mineral enthusiasts, especially since he knew almost nothing about that kind of operation. But invest he did, though only occasionally to any advantage.

That Ingram allowed himself the luxury of taking a chance in mining ventures in the early 1890's emphasized his own affluent circumstances and also the fact that investment opportunities were limited. Had conditions permitted, he would have preferred to continue investing in something he knew about, specifically in pine land. Indeed, he gave that matter long and hard consideration. Together with Major Day, Ingram carefully reviewed many possibilities for starting another Empire Lumber Company, either in the pineries of the South or on the Pacific slopes. The principal deterrent in the Pacific Northwest involved a question of market location. In the case of the southern alternative, although money was made in a number of separate ventures, the ideal situation just never seemed to come along.

Ingram was properly impressed by the size of the trees and the extent of the forests of Washington, Oregon, and northern California, but he was never quite satisfied where the customers for all the lumber would be found. As early as 1883, during one of Ingram's first journeys to the west coast, Colonel Dulany had informed Chamberlin that he had overheard the president of Empire "talking of digging up the Empire L. Co. and transporting it to California," even predicting the method of the transfer: "We will all go over in a Balloon."[6] But nothing of the sort developed.

After much deliberation, Ingram and Major Day did finally make some purchases in the Willapa valley, including a small sawmilling facility at South Bend, Washington, but neither was sufficiently enthused about the project to force any serious attempt at making the Pacific Empire Lumber Company an operational success. In fact, one has the impression that in the course of their travels across the country and their negotiations for timber and milling sites, Ingram and Day were both feeling a little old and tired, often acting more out of habit than from conviction. There obviously was not enough time left for them to build a new Empire in pine, and that may have been another reason that mining investments seemed to hold a special appeal.[7]

Still they held on to the South Bend property for some fifteen years, Ingram being far more interested in the possible start of operations there than was Major Day. Finally in 1909, following a Pacific Northwest vacation which included children and grandchildren, Ingram and Day agreed to sell their Washington lands to the Weyerhaeuser interests. This decision may have been prompted by the fact that the Major had lost his second wife, had recently married a young belle, and needed cash as a substitute for youth. Without doubt, Day had more on his mind than a sawmill in South Bend.[8]

Earlier in 1893, however, attention had quickly turned from a concern with investing to the best possible advantage to a fear that investments and savings of any sort might be lost. The effects of the panic, which began in February with the failure of the Philadelphia and Reading Railroad, became apparent in the upper Mississippi lumber markets early in the season. The

[6] D. M. Dulany, Sr., to Chamberlin, November 15, 1893.

[7] These are difficult attitudes to pin down, but Day occasionally provided some insights. In his letter to Ingram of September 4, 1900, for example, he made no attempt to hide his disappointment over the future of their investment at South Bend, Washington: "While I regard an investment in stumpage on the West coast if properly bought, as fairly promising for profitable returns, yet I do not think that the returns will be speedy and I am not personally so desirous of investment only expected to yield profit at some distant period of time."

[8] Charles Ingram of Tacoma, Washington, in conversation with the author, recalled many of the details of that 1909 trip with his grandfather, grandmother, Major Day, and others.

month of May proved particularly depressing to Ingram. Failure of the National Cordage Company emphasized the fact that even with the Democrats in control of Washington, the nation faced financial difficulties of no small order. Colonel Dulany could offer nothing in the way of encouragement from Hannibal. On May 8, 1893, he informed Chamberlain, "It is still raining here, no signs of letting up, farming prospects never more gloomy and the business outlook anything but bright." Although this was almost a carbon of correspondence from the previous year, for once the complaints were justified. To make matters worse, personal tragedy struck the Ingram family when Charley's wife, Grace, died after an illness of some duration. Ingram had been unusually close to this daughter-in-law and she to him, and he felt her loss severely.[9]

As the weeks passed, money which had been so recently an item of surplus in the Empire office and among the Empire stockholders again became scarce. While there may have been few organizations better prepared for the onset of a depression than the Empire Lumber Company, all of Ingram's business interests were not similarly situated. The "stringency" caused considerable hardship for such operations as the Hudson Saw Mill Company, and Ingram tried with little success to borrow money to help in tiding his associates over the difficult days. Fearing a run by depositors, banks were generally unwilling to lend even at high rates and low risk. The businessmen who needed money, if only to meet operational expenses, lowered prices in the hope of securing a larger share of the market. Even the Empire Com-

---

[9] The modern reader has difficulty imagining the number of personal tragedies which befell those who lived in earlier times. The Ingrams lost two of their six children as infants. They would lose two more as young adults. Small wonder that nearly every personal letter begins and ends with words of concern about the health of loved ones. Grace, known better as Kitty, seems to have been one of the few members of the family who made "Father Ingram" the object of their pleasantries, a talent he must have enjoyed. During a trip east in 1891, Kitty spent a few days in Tidioute, Pennsylvania, where she had opportunity to visit the J. L. Grandin family. Grandin, who amassed a fortune in oil and subsequently invested in numerous other interests including vast acreages of Dakota wheatland, had occasionally been an important source of credit to Ingram. Anyway, Kitty noted in a letter to her father-in-law on July 9, 1891, that the Grandins were "delightful" people who "spoke very highly of you." Then she observed with an obvious smile, "I guess you never told them you are a strong *democrat*," adding, "I think it is very wrong of you to deceive them. . . ."

pany, at the height of solvency, could not run indefinitely without receipts from lumber sales.

As inevitably happened during the periods of slow sales and slack demand, lumbermen began to talk about the advisability of shutting down the mills, thereby reducing expenses as well as reducing the amount of lumber on the market. During bad times, curtailment seemed the logical solution. Whenever lumbermen gathered, they would talk in glowing terms of the benefits which would be theirs if they could join together and produce less lumber this next season than had been produced the season past. Trade journals seldom printed an issue without calling to their readers' attention the common advantages of bringing the supply of lumber down to a level equal with demand. But the actual decision to close down early was made with great difficulty, for, as one lumberman observed, "This is a lumber country and we have no other employment. Men must eat to live and must work to eat. . . ."[10]

In the summer of 1893, talk of curtailment was resumed with vigor. As usual, however, an early end to production was wise policy only for other operators. Typical was Colonel Dulany's August letter to Ingram advising him to attend the Lumberman's Association meeting in Minneapolis that month where curtailment was to be the prime topic of discussion. The Colonel completely agreed that mills should be shut down "as early as practicable in order to prevent a perfect demoralization in prices for lumber." He urged Ingram to attend the August 8 meeting and support the drive toward an early end to the 1893 sawing season. But he quickly added, "in doing so we would want you to make the Empire Lumber Company an exception to the rule, for the reason . . . we got a very late start in sawing, then were stopped by high water, break downs and other troubles. . . ."[11] This was the reason that any consideration of curtailment among the independent producers of lumber amounted to "so much smoke," as Clarence Chamberlin so often described it. At the association meeting, fully confident that unanimity could not be achieved, Frederick Weyerhaeuser made his own views on the subject suf-

[10] *Northwestern Lumberman,* September 28, 1878.
[11] D. M. Dulany, Sr., to Ingram, August 5, 1893.

ficiently clear, stating: "Unless the others will shut down, we will run until it freezes up."[12]

Mills would indeed continue to saw logs into lumber until stopped by some natural catastrophe, or until it was clear beyond any doubt that it was to the owner's own advantage to close the doors. The seriousness of the depression which began in 1893 is attested to by the fact that many mills, for the first time in their long history of operation, did reach decisions to end seasons earlier than usual. The letter which Major Day addressed to the employees of the Standard Lumber Company in mid-September, 1896, was repeated in substance many times throughout the industry:[13]

> During the twenty years in which we have operated our mill at Dubuque, we have, almost without exception, opened with [the] first favorable weather in the Spring, and suspended operations only when compelled to do so by approach of Winter.
>
> We regret to state that, owing to the unfortunate business depression through which the country is passing, we will be obliged to close the mill this year about October 1st. But for our reluctance to throw out of employment those who have been depending on us for work, we would have ceased manufacturing earlier, as the demand for our product is exceedingly limited, and the prices realized are below the cost of production.

The complaints from Hannibal rang with an unfamiliar sincerity during the summer of 1894. For the first time the Colonel's direst predictions had been largely fulfilled, but he was unable to take much satisfaction from the fact. In July, as proof of his straitened circumstances, the old man told Ingram that his only pleasure those long days was to take a ride after supper out to his farm where he could observe "the effects of the great drouth and the ravages of the chinch bugs." Encouraged by the truly hard times, the Dulanys began to talk of closing down the operation. "As much as I may regret it," the Colonel wrote, "yet I must confess that it looks to me like the business had played out here."[14] Admittedly conditions in the lumber markets were generally poor; but among the Empire interests, Hannibal was far and away the poorest.

---

[12] Ulrich to Chamberlin, August 28, 1893.
[13] Major Day to "Our Employees," September 14, 1896.
[14] D. M. Dulany, Sr., to Ingram, July 23, 1894.

From Winona, Will Tearse reported to the Eau Claire office that he could find absolutely nothing "cheering to write" concerning the sale of lumber or prospects for the future. He added, no doubt for the benefit of Chamberlin, that he considered their misery in Winona to be entirely of their own making, and that since there had been no recent mail from Hannibal, this time it would be unfair to charge "anything in the blue line to them."[15] But although conditions were unfavorable at Winona, for the 1894 season that branch of Empire managed a small profit of $24,000, an amount equalled by Hannibal's losses during the same period.

As reported under the provisions of the new income-tax law for corporations, the gross receipts for the Empire Lumber Company in 1894 totalled $307,685.05. Through the magic of bookkeeping, the total operating and business expenses came to nearly that same amount. In any event, and by all previous standards, 1894 had been a very bad year, and as Ingram carefully reviewed the books he grew more receptive to the thought of bringing operations in Hannibal to a gradual end. There needed to be nothing impulsive done. Indeed it would be to the advantage of all concerned if the process of closing out the wholesale yard were accomplished in a slow and orderly fashion. The logical time to stop producing lumber at the Empire mills in Eau Claire was when Empire stopped marketing lumber in Hannibal. A gradual cessation would permit less painful adjustments on the part of those involved in the logging and sawmilling operations. Also, a much greater return on the Hannibal inventory could be assured if sales were not forced and if the assortment remained complete as it was being reduced. Ingram, therefore, declined to push the question. From a personal view, he was in no particular hurry.

While willing to accept the approaching end to operations at Hannibal, the president of Empire maintained the hope that they would be able to continue in the business until the pine in the valley was gone. As a final example of their differing attitudes, if the Dulanys feared that depressed conditions had become a permanent fact and that accordingly their wholesaling efforts at Hannibal ought to be concluded without delay, Ingram

---

[15] Tearse to Chamberlin, July 24, 1894.

thought more in terms of a sudden improvement in the financial situation. If he had any fears for the future, they tended to focus on the possibility of customers hungry for lumber and Empire yards with nothing to sell. He knew the depression soon had to end and, with the return of good demand and high prices, the Dulanys would enjoy a revival in business activity and in spirits from which they would all stand to profit. At any rate, it was clearly not a time for impetuosity. The decision to end production at Eau Claire would be made only after the most careful deliberation. Once made, there was no possibility that the decision could be reversed.

In spite of Ingram's stubborn optimism, the depression wore on and on. The slight upswing enjoyed in the fall of 1895 was more seasonal than symptomatic of any lasting improvement. Prices soon fell again, and to levels lower than had previously been the case in the lumber markets. In early March, 1896, Colonel Dulany expressed the opinion that they were about to enter what would prove to be "the darkest year for business we have had." With the unsettled political situation and the general feeling of depression across the land, the only thing he saw on the horizon which might result in some change was war with Spain, which old Dan believed "would be unfortunate, just from a monied stand point."[16]

But Colonel Dulany was no longer unusual in his prophecies of gloom. Perhaps no better evidence exists as to the depths of demoralization among lumbermen than the proceedings of the annual meeting of the Chippewa Logging Company, held at Chippewa Falls on September 9, 1896. In addition to Weyerhaeuser and Ingram, those in attendance included William Carson, Delos R. Moon, Sumner T. McKnight, David Dubach, William H. Laird, E. S. Youmans, Thomas Irvine, H. G. Chichester, and Clarence Chamberlin. Traditionally one of the more crucial discussions at such meetings involved the determination of the quantity of logs which the Pool would put in during the coming winter. As the initial step in this determination, individual members indicated their estimated needs. Then, after considering a variety of other factors, a total figure was approved and the distribution to member mills followed the simple formula based

[16] D. M. Dulany, Sr., to Ingram, March 5, 1896.

on percentage of ownership. Because these deliberations held such basic importance to the participating firms, the officers in attendance generally debated the subject with considerable spirit. In 1896, however, that was not the case. In response to the question of how many board feet of logs were wanted for the 1897 season, only Ingram and Weyerhaeuser replied with specific figures, both indicating "50 to 75 Million," although Weyerhaeuser implied a preference for the lesser amount. The others answered as follows: "Mr. Moon – 'None'; Mr. Youmans – 'Fewer the better'; Mr. Irvine – 'None'; Mr. Dubach – 'None, only as necessary to save timber'; Mr. Laird – 'None, only as necessary to save timber'; Mr. Carson – 'Don't want many.' " Good times were recalled with increasing difficulty.

Lacking any tangible evidence of improvement in financial prospects, Ingram could not insist indefinitely that others continue to share his optimism. In one of his last letters to Eau Claire, Colonel Dulany had reminded friend Ingram of "an old adage, and a true one, that there is an end to all things, sooner or later."[17] Dan Dulany died on July 2, 1897. At the annual meeting of Empire on September 13, 1898, the directors of the Company unanimously passed a resolution "discontinuing the manufacture of lumber at Eau Claire, Wisconsin, and authorizing the sale of the property of the Company at Eau Claire. . . ." All of the logs which had been received at the Eau Claire Empire mills under the distribution policy of the Chippewa Logging Company would in the future be delivered to the booms of the Winona Empire mill.[18]

As Ingram had been predicting for so long a time, conditions in the lumber markets finally improved during 1899, the season following the decision to stop production in Eau Claire. At many points, prices on common lumber advanced upwards of $3 per thousand feet. The Rice Lake Lumber Company, for example, sold three-quarters of a million feet less lumber during the 1899 season than had been the case in 1898, but receipts totalled nearly $100,000 more in 1899.

From Hannibal George Dulany agreed with Chamberlin that "we are very unfortunate in just going out of business at this

[17] D. M. Dulany, Sr., to Ingram, December 9, 1895.
[18] Bill of sale for logs, Empire of Eau Claire to Empire of Winona, April 27, 1898.

time, after having pulled through the dull period."[19] He might have added, however, that it was a very favorable period to be closing out the stock in the yard, there being no lack of customers willing to accept less than perfect articles or orders filled not exactly to specifications. By the first of November, most of the lumber had been sold and the remaining Dulanys moved from the yard to an office above an old clothing store. George described their new situation as being "well lighted and quite convenient" since it was directly across the street from the bank, although, he admitted to Chamberlin, "how much business we do is another question."[20]

By mid-summer, George had tired of talking politics day after day and took his family on a European tour. T. G. Dulany left the heat of Hannibal for the mountains of Colorado, leaving eighty-two-year-old William H., "the only Dulany in town, with the bag to hold." The elderly William, who had never seemed too happy when Empire at Hannibal was a lively operation, seemed even less so now. He acknowledged as much to Ingram: "In my active business life we never had time from pressing business to take outings or go pleasure hunting. But time has brought great changes, whether for the better is a question with two sides to it."[21]

Characteristically, there was no special drama in the determination of the Empire directors to cease operations at Eau Claire and Hannibal in 1898. It was simply another in a long series of business decisions, Ingram and his associates agreeing that the time had come to stop producing lumber at the old mills. An entire forest had been cut, banked, driven, and sawed into lumber since Ingram first viewed the pine of the Chippewa valley forty years previous; but unlike W. H. Dulany, he was still too busy to give the matter much reflection.

Ingram would agree with the oldest living Dulany, however, that the passing years had witnessed great changes in the lumber industry and in their own lives. Certainly the decision to shut down the Empire mills in Eau Claire significantly changed Ingram's personal activities. In effect, he suddenly ceased to be a

[19] George W. Dulany to Chamberlin, February 24, 1899.
[20] George W. Dulany to Chamberlin, March 24, 1900.
[21] W. H. Dulany to Ingram, August 3, 1900.

manager. The Winona and Dubuque operations had long been essentially separate and self-supporting. Ingram could hardly involve himself to a larger extent at either point without running the danger of interfering with established procedures and the capable management of Will Tearse and Major Day.

Rice Lake was another matter. There the managers were neither family nor friends of long-standing and Ingram felt less constrained about providing personal direction, and direct he did. In fact, nothing provided more pleasure to him in his later years than watching over the growth and success of the Rice Lake Company. There are no doubt a number of reasons why this was true. In the first place, it was a neat and all-inclusive operation and one that had been solely his idea. Initially this investment had attracted little support from any of his Empire associates. Furthermore, because it was separate and situated in another river valley, one heretofore monopolized by the Knapp-Stout organization, it was necessary to build it from scratch and this effort occurred at a time when Ingram could provide all necessary credit. No penny-pinching at Rice Lake—if new saws were needed, new saws were installed; if night operations seemed desirable, the plant would be wired for incandescent lighting. The product of the Rice Lake Company was shipped to markets entirely by rail, something Eau Claire had rarely done. But times were changing, and clean, white, finished lumber was in greater demand than had been the rule when board fencing and barns, sheds and cribs provided the market basis.

In some respects, Ingram's attitude towards the established Rice Lake Company was similar to that of an industrialist who was a gentleman-farmer. Not that Rice Lake would fail to pay its own way. It was a very successful operation, at least as concerned pine lumber, producing millions of feet and selling in nearly every state at considerable profit. But Ingram's involvement at Rice Lake could be and often was a sometime thing, especially when the operation there was running smoothly. Yet it offered him the opportunity to keep abreast of what was new, and the Rice Lake sawmill was filled with the latest in machinery. As a final but important advantage, Rice Lake was convenient to the Ingram summer homes on Long Lake and Lake Sissabagama. There was no better way to enjoy oneself than to fish until the conscience bothered, whereupon "O. H." could take a quick run to Rice Lake. There the work went on, but one might always offer suggestions—even if only about the water levels in the fire barrels.

Yet in no sense was Ingram retired. His responsibilities following the end of the 1898 sawing season were simply of a very different nature than previously. Although his interest remained more than just casual, his relations with the Winona and Dubuque operations became the pattern of his future involvement in lumbering, and in other areas as well. Active and direct participation in the actual functions of a business gradually lessened as Ingram became, to all intents and purposes, an investor.

Whether he fully realized or cared to admit it, Ingram had clearly reached the point where he could allow his money to do the work while he shortened his own hours at the office and began to sit back, relax, and reap the benefits of long years of hard labor. The *Biographical Volume* published by the *American Lumberman* in 1905 included an observation that "large business ventures have not kept Mr. Ingram from enjoying life as he went along." That would have been nice had it been true, as it must have been true for some. In Ingram's case, however, it was the "large business ventures" that he most enjoyed. In 1912, at the age of eighty-two, Ingram admitted that he was anxious to rid himself of some interests that were "causing me more work than I ought to do at my time of life," but he quickly added: "I shall want to do something as long as I have strength, for I would not be happy without participation in business."[22] For individuals like Orrin Ingram, life without business involvements and business dealings was no life at all.

He and his family were fully equipped to take advantage of opportunities for increased leisure time. He owned a small farm on the outskirts of Eau Claire where he kept fine horses and raised a few pigs and some turkeys. At the comfortable cottages on Long Lake and Lake Sissabagama, seventy miles to the north, he often entertained and would spend hours trolling from the stern of his steam launch. *The Fannie Ingram*. Actually he was much more the dedicated fisherman than might be expected, although one cannot help but wonder whether he thought more of business than of pleasure while indulging in this pastime. No

---

[22] Ingram, *Autobiography*, 81.

casting was permitted from the boat and to an observer the activity must have seemed as passive as it was peaceful, especially if the fish were not biting. As noted, Ingram enjoyed an occasional cigar, but when fishing he preferred a bit of chewing tobacco which he took from a little silver case. While chewing and concentrating on the bobber of his cane pole (a metal pole would be substituted when muskies were the quarry), now and then a little juice would escape to stain his pure white chin whiskers, lending a hint of casualness to an otherwise stuffy sportsman, all hunched over in business suit and tie. But his greatest pleasure remained as before, going to the office. From 1895 Ingram's business headquarters were located in a magnificent Romanesque structure in downtown Eau Claire which he had constructed in accord with his own ideas of what a modern office building ought to be. It was named, after the custom of the day, the Ingram Block.

Although the depth of his participation in any one business activity had lessened considerably, Ingram compensated in large measure by increasing the number of his involvements. He was president of the Eau Claire National Bank until January, 1905, when he retired, only to be appointed president of the successor organization, the Union National Bank, in 1911. This was one of the interests which he retained with little enthusiasm. In 1894 he had purchased the Eau Claire Pulp & Paper Company for $48,500, outbidding others at the receiver's sale. He had no real interest in the paper business, but that company owed his bank nearly $50,000 and he was unwilling to take a chance on suffering the loss of such an amount. He soon persuaded the Davis brothers, experienced paper makers of the Neenah-Menasha area, to come to Eau Claire and assume management of the new Dells Paper & Pulp Company. David R. Davis subsequently became the president of the organization, with Ingram serving as vice-president, Clarence Chamberlin as secretary, and Charley Ingram as treasurer. In 1900 the Ingram interest was sold to the Davis family, and by 1914 the Dells Paper & Pulp Company was the largest single industry in Eau Claire.

For a time Ingram served as president of the Eau Claire Water Works Company. Later he was president of the Chippewa Valley Light & Power Company, capitalized at $1,265,000, and president of the Chippewa Valley Construction Company, both of which were incorporated in 1897. Interests such as these were a natural outgrowth of his long involvement in Eau Claire. Less easily explained in his growing list of investments were undertakings such as Florida orange groves, a Texas rice plantation, and a copper mine in Arizona. There always seemed to be a special reason, often not strictly business, for each separate interest. He was assisting an old associate in the rice plantation. In the case of the copper mine, a son of his old friend and former partner Donald Kennedy had overextended himself financially and Ingram wanted to provide assistance and possibly mine a little copper in the process. He and Charles Bullen of the Daniel Shaw Lumber Company shared the interest in Florida oranges for no obvious reason, although one suspects that Ingram would have enjoyed spending vacations in Florida, a possibility which would become more conscionable with an excuse of business thereabouts. Ingram was simply not the sort who could justify to himself, and therefore enjoy, a vacation for its own sake, even in sunny Florida.

Of much greater importance, both as to money invested and time demanded, was the Canadian Anthracite Coal Company. Located high in the Canadian Rockies, it came as no great surprise that mining was less a problem than marketing, and dividends were not taken for granted. Both Ingram and Weyerhaeuser were major stockholders and directors of the coal company, but of the two Ingram had the greater involvement. He served as treasurer for many years, an often demanding but nonetheless satisfying responsibility. Not only was Canmore, situated just east of Banff in the Bow River Valley, an exquisite setting for an investment, Canadian Anthracite proved to be a steady payer, returning an average dividend of about 10 per cent a year.[23]

Coal mines and the like were, however, in the words of Colonel Dulany, strictly "side show" operations. For all of the diversification apparent in Ingram's investments beginning in the 1890's, the others remained relatively unimportant when compared to his continued interest and involvement in the lumber industry. In addition to the presidency of the Empire Lumber Company,

[23] Canmore was the name given to the operating company of Canadian Anthracite. Production continues there, and the Canmore mine was the first in western Canada to ship coal to Japan in 1959 through Portland. Recently merged with the Dillingham Corporation, the mine now ships some 80 per cent of its product to Japan via Vancouver.

he was also president of the New Dells Lumber Company, president of the Rice Lake Lumber Company, president of the Wabasha Lumber Company, president of the Fort Scott (Kansas) Lumber Company, vice-president of the Chippewa Lumber & Boom Company, vice-president of the Standard Lumber Company, and a director in the Chippewa Logging Company, the Hudson Saw Mill Company, the Hannibal Saw Mill Company, and the Woodville Lumber Company. But by the end of the first decade of the twentieth century, most of these lumber concerns were in the process of closing down and liquidating their assets, if they had not already done so. There were other forests, however, and other fortunes were possible for those engaged in the conversion of trees into lumber, even if the lumber proved a lesser quality than the familiar white pine of the Lake States region.

Although Ingram never expressed much enthusiasm over prospects in the southern pineries, he finally made some investments in that section, the first of which involved the Louisiana Long Leaf Lumber Company, more commonly designated the 4 L Company, with headquarters at Fisher, Louisiana. He later invested in the Louisiana Central Lumber Company of Clark, Louisiana, and added partial interests in the White-Grandin and the Louisiana Saw Mill Lumber companies. Collectively these four operations came to be known as the White Mills, as his old friend Captain J. B. White of Kansas City had been instrumental in their establishment. Although Ingram took no active role in the White Mills, they proved to be very profitable investments.

Such was not the case with the Ingram-Day Lumber Company of Lyman, Mississippi. This partnership, capitalized at about $500,000, had resulted only after a considerable search for the ideal situation in the southern pineries. But the conditions never quite proved out, perhaps because the elderly partners never quite approved of the conditions. Laborers were different; logging methods were different; and the yellow pine lumber was, as they had long known, very different from that they had produced in northern mills. Apparently the greatest single problem, however, was to obtain a manager in whom both Ingram and Major Day could have complete confidence. For whatever reason, the Ingram-Day Lumber Company never got such a manager. In 1912 Ingram readily admitted that Ingram-Day had never been a paying proposition, adding, "If I should get my money back without interest I shall be satisfied."[24] Doubtless one of the reasons for Ingram-Day's poor showing was the reluctance of the owners to make the necessary sacrifices. Major Day and Ingram tended to forget how hard they had worked and with what single-purposeness to achieve success at Dubuque and Eau Claire. But these past achievements could not assure success at Lyman, Mississippi, or on the Pacific slopes.

Ingram and Day's co-operative undertaking in the Willapa valley of Washington was pursued with even less commitment than their southern venture. Just how far Ingram might have been willing to go in the matter of land purchases and the construction of sawmilling facilities for the Pacific Empire Lumber Company is not clear. Certainly he was far more ready to increase his investment than was the Major. One of the complications in these efforts to establish a new lumbering operation involved a disagreement over which area held the greater promise, the Major preferring investment in the South while Ingram inclined toward the Pacific Northwest. But in truth, neither was sufficiently dedicated to force the issue or to break away and proceed alone or in concern with new associates.

Without any direct admission, this reluctance to start anew reflected the nature of Ingram's new relationship with the lumbering industry. The time had come for others to travel through the woods, superintend the details of the sawing operation, and to worry over the arrangements for marketing the product. No doubt, in the eyes of an aging Ingram, Frederick Weyerhaeuser was the successful one. To others, however, Ingram's own success was notable.

As a teenager he had been introduced to the business of making lumber at the small mill near Lake George. There was little in the process he had not experienced at one time or another. It became early apparent that the young Ingram had a definite talent as a craftsman, one skilled in designing, building, repairing, and improving the machinery in the sawmill. In the pine regions of Ontario and the Ottawa valley he had unusual opportunity to develop these skills and to acquire new ones. He designed not only machinery but entire mills. He directed their

---

[24] Ingram, *Autobiography*, 81.

construction, staying to supervise the sawing, thereby learning the practical results of his innovations. Working for others, he began to understand something of the problem of handling men and quickly appreciated the satisfactions that came with making money. Like so many others, he eagerly awaited an opportunity to go into business for himself.

In some respects, it was largely accidental that the Chippewa valley of Wisconsin proved to be the locale for Ingram's independent effort, but in retrospect the choice appears to have been quite logical. By the late 1850's, the lumbering frontier was poised for its assault on the Wisconsin pineries; and for those with limited resources looking for unlimited opportunities in lumbering, there was no better selection possible than the one made by Dole, Ingram, and Kennedy. There would always be problems of a natural sort requiring solution, but the Eau Claire situation offered advantages second to none.

Through the Civil War decade, the period of mill construction and improvement of facilities in Eau Claire, the attention of Ingram and his partners focused primarily on the production of quality lumber in the most efficient manner possible. During the 1870's, increasing competition in the markets of the upper Mississippi valley made a concern with marketing procedures, particularly the establishment of a dependable network of wholesale yards, a subject of priority. In the 1880's, the decreasing availability of forest resources forced Ingram to turn his energies to the procurement of sawlogs for the mills of Empire. From craftsman, to producer, to manager, to investor, Ingram's career reflected a fundamental process of change in the lumbering industry and, to an extent at least, in the more general history of American industry in the last half of the nineteenth century.

Without question the most important decision of Ingram the investor was made in June, 1900, when he decided to purchase 2,160 shares in the newly organized Weyerhaeuser Timber Company at an initial cost of $216,000. The purchase is an interesting one from a number of considerations. In spite of his own reluctance to expand the activities of the Pacific Empire Lumber Company, Ingram apparently had reconciled the problems earlier foreseen regarding the marketing of lumber produced in the Pacific Northwest. Furthermore this was an investment in which Ingram would have little to say about the actual management procedures and policies. Having a say, however, may not have been quite as important as in years past.

The Rice Lake Lumber Company, and even the construction of the Winona sawmill, had special appeal as a result of the greater independence which these operations seemed to offer. But, as Major Day had observed in 1891, it was not just on the Chippewa where one could see "the broad German hand of Mr. Warehauser [*sic*] pulling the wires. . . ."[25] For all of the quiet complaining, however, Weyerhaeuser had proved beyond any doubt his ability and wisdom as a lumber industrialist. Whatever other feelings might have occurred to Ingram at the time, he knew that there could be no wiser investment than stock in the Weyerhaeuser Timber Company, then just beginning to purchase the choice timber and mill sites in the state of Washington.

Although Ingram kept himself reasonably well-informed of the progress of the Weyerhaeuser company, he chose to remain in Eau Claire and seldom journeyed west to the scene of activity in the woods and mills. He left the actual participation to his sons and grandsons. Indeed, the Weyerhaeuser Timber Company stock had been purchased in his son Charles' name, but after an extended illness, Charley died in 1906 at the age of forty-seven.

Together with his remaining son, Erskine B. Ingram, and his son-in-law, Dr. Edmund S. Hayes, Ingram then organized the O. H. Ingram Company, largely for the purpose of managing the many and diverse investments of the Ingram family. Although Erskine was a meticulous bookkeeper, he abhorred the business discussions and decisions that were so enjoyable to his father. Fortunately, Dr. Hayes had a more natural bent for such affairs and he came to be a very close friend to Ingram in the old gentleman's later years. Edmund quit the medical profession in order to assume greater responsibilities within the Ingram Company, and if he could not share completely his father-in-law's enthusiasm for the many interests and investments, he did prove to be a capable and deliberative manager, one in whom Uncle Henry had absolute trust. Following each Sunday meal at the Hayes's home, after the dishes were cleared, Dr. Hayes and Ingram would retire to the study where they would fill the room with cigar smoke and business talk. In truth, however, neither Erskine nor

[25] Day to Ingram, May 18, 1891.

Edmund were much interested in lumbering. Those involvements skipped a generation, and the major contributions from the family to the management of the new Weyerhaeuser interests in the Northwest came from grandsons Charles H. Ingram and Edmund Hayes.[26]

There is no question but that Ingram enjoyed far fewer satisfactions in business after the closing down of the Empire mills in Eau Claire, at least in the sense of any direct participation. He made up for this in part by increasing his involvements in local affairs, and in what he would have called good works. This was not merely an attempt to build his own local monuments. His concern for the community was sincere, a concern that was not shared by all resident lumbermen. Joseph Thorp, for example, spent no more time along the Chippewa than was absolutely necessary for purposes of doing business there. Many doubtless shared the attitudes of Delos R. Moon who, in recommending to Clarence Chamberlin that they send their daughters off to Europe for a year, observed, "There isn't much for young girls that is good to be gotten out of Eau Claire & much less for young boys."[27] Ingram wanted to make Eau Claire a better place to live, for young and old alike.

He took a special interest in the Young Men's Christian Association, an organization which had been very active in the Wisconsin pineries during the winter months, bringing sermon and song to willing and unwilling loggers alike. Ingram gave $20,000 towards the construction of a Y.M.C.A. building in Eau Claire. In early 1895, after the completion of the Ingram Block, he donated the use of a large part of the first floor to the Eau Claire Public Library, the rental of which would otherwise have cost several thousand dollars a year. He contributed a thousand dollars towards the completion of the local Children's Home and a substantial amount to the Eau Claire Sanatorium.

Not unexpectedly, as Ingram became less actively involved in the affairs of business, his family assumed an increasing place in his life. But even at home he seems to have had difficulty relaxing and enjoying in any complete sense. Henry, as Cornelia

[26] Hidy, *et al.*, *Timber and Men*, 278, 328, 333, 402, 423, 436, 502, 508, 554, 577, 633–634.
[27] D. R. Moon to Chamberlin, February 1, 1895.

Ingram called him, could hardly qualify as the doting grandfather. Indeed, his own grandchildren seldom acknowledged their relationship, addressing him as "Mr. Ingram," and when they spoke, it was at the old man's invitation. Perhaps he was simply not at ease with children. In any event, outward expressions of warmth did not come easily to Ingram and his bearing and manner were such that youngsters naturally kept their distance and their place.

For the most part, the passing years had been kind to Ingram. He remained a figure of strength, easily carrying more than two hundred pounds on his large frame. His pure white hair and beard were carefully groomed, and if age had slowed his step and made hearing more difficult, his cane and ear trumpet were accouterments which somehow added to his distinction. Few were unable to recognize the "old gentleman" from a long ways off. But grandchildren can get accustomed to appearances, even to ear trumpets. It was "Mr. Ingram's" demeanor that regulated their conduct. Perhaps he merely wanted to set the best of examples, and the results of his own good life were obvious enough. The children heard no profanity from him and would not have dared such expressive experiments in his presence. He also knew that foolishness need not be encouraged. Seriousness and strength, piety and frugality: these were the qualities which would prove important in the long run—and the sooner young people understood, the greater the chances for their success. In other words, the sooner children ceased being children, the better it would be for all. Not surprisingly, then, dinner at the Ingram table was a quiet, business-like experience. One took satisfaction from the food and perhaps from the fact that Grandfather Ingram's prayers, always ending with the intonation, "Bless this food prepared for our nourishment," were relatively short. Such things are important to youngsters.

Ingram did not miss an opportunity to assist in the proper development of his children and grandchildren. Thus the following note was addressed to grandson Charles on December 25, 1915: "I have thought best to give to each one of my grandchildren a five dollar gold piece and ask them to buy something they would like to have as a Christmas and New Year greeting from me, or, if they prefer, deposit it in the savings bank where it will increase in value." Even without supporting evidence, one has the definite feeling that on Monday, December 27, a number of five-dollar deposits were accepted by the local savings bank.

In spite of none-too-subtle pressures and the general severity of the relationship, the younger members of the family did enjoy the company of their grandfather. Fishing proved to be the greatest bond, an activity which all manner of types could share. The old man and the children could be together in the boat and far apart in thought; conversation was seldom necessary. Of course, Ingram was interested in what the youngsters caught, but not to the detriment of his own efforts. Typical was an experience which Charles had with his grandfather while fishing on Long Lake, the site of the Ingram summer home. On this particular day the fishing had been poor and rain had begun in earnest, but Mr. Ingram refused to notice. The guide, then called a "rower," was a Swede of long employ, a man by the name of Seaver Soholt. Seaver would bait the lines, take off the fish and generally keep quiet, but in this instance, worrying about the boy's getting wet, Seaver suggested that it was time to be heading for shore. Ingram replied that they would go in "when we catch one more fish." Soon afterwards young Charlie landed a small bass and announced, "Now we can go in," to which his grandfather replied: "We will go in when *I* catch one more fish."

A further bond between the oldest and the young Ingrams had to do with what might properly be called a weakness on the part of "Henry." For all of his moderate if not severe qualities, grandfather was a youngster in the sense that he was forever interested in the latest, the most recent. When horses were the only option, his matched pairs were among the finest to be seen on the streets of Eau Claire. But when the horseless carriage became available, only the most up-to-date models carried Ingram and his family about. Whether it was the White Steamer, the Locomobiles, or the White gasoline automobile, the grandchildren found much to fascinate, and to scrub and polish. Not even the fishing was unaffected by technological change, Ingram purchasing one of the area's first outboard motors. This he mounted on a large box-like hull, some ten feet wide, which he thought would prove ideal for trolling Long Lake—one could walk around and thereby not get stiff. But like so many early models, performance failed to live up to promise. The motor started with great difficulty and ran poorly thereafter; soon the amused grandchildren christened the vessel "Cleopatra's barge."

Aside from family affairs, social life for the Ingrams was and always had been limited almost exclusively to the church. Indeed, those who found pleasures elsewhere were considered worldly if not downright sinful. Henry and Cornelia Ingram kept track of days and friends through the doors of the building next door, the Eau Claire Congregational Church. Sundays, of course, began with church school and the morning services. As the years went by, Ingram found it necessary to sit closer and closer to the front until finally he was in the very first pew, leaning forward, ear trumpet affixed. There were a variety of other activities through the week: recitals, sermons, or lectures by visiting dignitaries and discussion groups. And each Friday afternoon the women would troop to the church basement "to butter the bread," this in preparation for the Friday evening social which featured doughnuts, beans, brown bread, and coffee, all for fifteen cents. Once a year, the Ladies Social Circle would put on an oyster stew, and for this special event the charge was a quarter.

Needless to say, Cornelia would permit only problems of health to interfere with her contributions of time and energy. If Ingram was a bit less devout than his wife, and his occasional cigars and private sips of Madeira were clearly evidence of weakness, he nonetheless was a most important member of the Church. For many years he served on the state board of trustees for the Congregational Church, but to the Eau Claire congregation his most important contributions were local and financial. Even his wife must have approved of the practical expression of his support. Every year at the annual meeting, the financial report would conclude by noting the annual deficit, and every year O. H. Ingram would arise to announce that he would accept personal responsibility for making up that amount.

But the Ingrams did not limit their interest and support to the Congregationalists. Together they were interested in and contributed to many different causes and denominations. Following the death of their son Charles, they gave $40,000 for the construction of a small memorial church in Washington, D. C., having made a similar gift toward a church in Boise, Idaho, when their daughter Fanny died in 1895. In the second decade of the twentieth century, when fire destroyed the First Congregational

Church in Eau Claire, Ingram contributed $30,000 towards the new building. He was a warm admirer of the famous evangelist Dwight L. Moody, and gave regularly to Moody's Theological Seminary in Chicago, served on the board of the American Board of Missions, and supported a variety of foreign missionary efforts. Few mailing lists to philanthropists appear to have overlooked the president of the Empire Lumber Company.

As a trustee of Ripon College, Ingram took his responsibilities very seriously and became intimately involved in the plans and problems of the little school. The institution benefited considerably from his interest, loyalty and generosity, as evidenced by a new science building on campus, Ingram Hall.[28] Without question the fact of such an edifice had real appeal to Ingram, but this does not explain his respect for Ripon and for higher education in general. In 1894 he sat down and wrote a letter to his nephew, Orrin H. Ingram, Jr., whose education at Ripon the elder Orrin was financing. Having learned that the young man was considering quitting school, Uncle Henry offered advice:[29]

> You cannot imagine now how much better fitted a man is for business when he has the advantage of a college education. I can realize that because I have been deprived of that; and have met so many times, with men in business who have had a college education and with those who have not. And my observation is that every young man who can possibly arrange to have the advantages of a college training should avail themselves of it without mistake. . . .

Ingram was also chairman of the State Capitol Building Commission, and in 1899 he accepted the less pleasant responsibility of chairing the State Relief Committee, which had been organized following one of the worst disasters in Wisconsin history, the tornado which struck the little community of New Richmond on the evening of June 12. More than a hundred people were killed in that brief storm, hundreds more were injured, and property

---

[28] Science halls seem to have been a special weakness for lumbermen supporters of colleges. For example and in addition to Ingram and Ripon, Weyerhaeuser contributed such a building at Macalester; William Laird of the Laird-Norton Lumber Company did likewise for Carleton College in Northfield, Minnesota; and Isaac Stephenson did the same at Lawrence College (now University), in Appleton, Wisconsin.

[29] Ingram to O. H. Ingram, Jr., April 16, 1894.

damage totalled nearly $800,000. Ingram never worked harder for himself than he did in the summer of 1899 for the unfortunate citizens of New Richmond.

This was the Orrin H. Ingram that most people remembered on the morning of October 16, 1918, when news of his death began to circulate around Eau Claire, a community just barely functioning in the midst of the influenza epidemic. Lumbering was already a subject for history along the Chippewa. If the townspeople were generally aware that the tall, straight old man with the white goatee had made his fortune in the white pine that used to grow in the forests of the valley, few remained who had shared the lumbering experience with him. Thorp and Carson had passed away many years before. Eb Playter had died in 1906, the same year as Charley Ingram. Friend Chamberlin and Cornelia Ingram both died in 1911, and Weyerhaeuser three years later. Brother Julius was eight-five when he passed away after a short illness in 1917.

Orrin Ingram had outlived them all; and, of greater significance, he had outlived the forest. In his sixty years in the valley of the Chippewa, few played a more important role than he in the exploitation of the timber resource. In years to come, much would be written about the destruction of the great forest; and perhaps too much would be forgotten about the processes involved, and the cheap lumber produced, and where it went, and what it did for an aggressive, westward-expanding America.

APPENDIX

TABLE I

LOGS AND LUMBER TRANSPORTED VIA THE
CHIPPEWA RIVER, 1878–1900[a]

Year	Lumber (Board Feet)[b]	Total Tonnage[c]	Loose Logs
1878	154,119,860		
1879	248,932,000		250,000,000
1880	325,150,000		300,000,000
1881	342,887,000		300,000,000
1882	375,000,000		350,000,000
1883	269,094,203		450,000,000
1884	298,344,591		534,674,176
1885	374,138,443		600,000,000
1886	207,205,672		465,000,000
1887	186,826,521		404,302,650
1888	161,309,512	325,971	542,437,000
1889	158,938,294	329,156	400,518,720
1890	166,477,966	342,350	606,992,790
1891	152,040,386	314,085	284,113,430
1892	144,651,150	410,960	632,350,670
1893	159,180,534	459,505	488,926,000
1894	128,703,908	333,157	250,045,730
1895	130,117,213	327,550	292,640,000
1896	104,715,810	258,776	297,391,000
1897	98,394,313	250,089	203,598,960
1898	109,012,223	301,842	174,964,700
1899	99,145,418	250,603	223,694,430
1900	90,019,811	232,079	202,697,860

[a] U.S. Army Corps of Engineers, Chief of Engineers, *Annual Report*, 1901–1902, part 3, p. 2328, in 57 Cong., 1 sess., House Documents, vol. 14, no. 2, serial 4281.
[b] Not including lumber which was shipped by rail from milling points on the Chippewa River.
[c] Not including loose logs, but including lumber, lath, shingles, and pickets.

TABLE II

DISTRIBUTION OF CHIPPEWA LOGGING COMPANY LOGS
TO NOVEMBER 30, 1886

Member Firm	Logs	Total Feet	Average per Log
Empire	852,474	165,112,750	193.68
Meridean	644,523	126,331,310	196.00
Valley	530,845	103,302,650	194.60
Northwestern	561,272	108,743,710	193.74
Badger	339,216	66,045,140	194.69
Westville	173,407	33,329,470	192.20
Dells	222,347	43,171,310	194.16
Total for Eau Claire members	3,324,084	646,036,340	194.35
M. R. L. Co.	8,418,301	1,621,987,920	192.67
TOTAL	11,742,385	2,268,024,260	193.148

TABLE III

COMPARISON OF LOG PRICES WITH WHOLESALE
LUMBER PRICES, 1882–1890

Year	Log Price[a]	Total Lumber Sales by Empire of Eau Claire	Lumber Price[b]
1882	$8.24	46,709,591	$13.025
1883	8.25	35,750,432	12.85
1884	8.08	31,849,554	12.04
1885	7.76	30,349,323	11.14
1886	7.69	32,242,340	11.30
1887	7.86	20,604,892[c]	12.84
1888	8.42	25,918,255	12.906
1889	8.49	25,561,047	12.99
1890	8.49	26,694,638	12.00

[a] *I.e.,* the price set by the Chippewa Logging Company.

[b] *I.e.,* the average price per thousand feet, wholesale, received by the Empire Lumber Company of Eau Claire.

[c] This sharp drop in Eau Claire lumber sales was not reflected in the total sales for the Empire Lumber Company. The 1887 season was the first in which an Empire Company sawmill was operated at Winona, Minnesota.

SELECTED BIBLIOGRAPHY

MANUSCRIPTS AND UNPUBLISHED MATERIALS

THE BASIC SOURCE for this study was, of course, the Orrin Henry Ingram Papers. This large collection of correspondence and business records is the property of the State Historical Society of Wisconsin, and is currently housed in the Eau Claire Area Research Center in the University of Wisconsin–Eau Claire Library. The letters touch on all facets of the lumber industry, although outgoing correspondence, in the form of letterbook press copies, does not begin until the middle 1870's. Without doubt, the Ingram Papers are one of the most important primary sources on the subject of Lake States lumbering in the nineteenth century. The dates of the correspondence are from 1857, the year of Ingram's arrival in Eau Claire, to 1904.

Also consulted with care were the Daniel Shaw Lumber Company Papers and the William W. Bartlett Collection, both of which belong to the Eau Claire Public Library. The Shaw correspondence is particularly useful in gaining some insight into the operation and special problems of a small and forever-struggling firm. Most of the outgoing correspondence consists of letters written by Eugene Shaw, son of founder Daniel. In addition to information on lumbering activities, anyone interested in fishing lore and entertaining stories would enjoy perusing the letters of Eugene Shaw. The Bartlett Collection comprises letters, recollections, business records, and materials of a general nature pertaining to the history of Eau Claire and vicinity. Bartlett dedicated much of his life to the collection and organization of items of local historical interest and significance. One is often surprised at what has been included, for example a large collection of excellent photographs, many of which deal with lumbering activities.

In the library of the State Historical Society of Wisconsin are the Marshall Cousins Scrapbooks. These consist of newspaper articles about the early residents of the Eau Claire area which have been clipped and pasted without the details of publication. The Scrapbooks are nevertheless invaluable as a point from which to begin any biographical investigation of the Chippewa valley. Each volume is indexed. Also available in the Society library is a scrapbook of articles by Captain Fred A. Bill, entitled "Navigation on the Chippewa River in Wisconsin." These articles originally appeared in serial form in the *Burlington* (Iowa) *Post* in 1930.

The Joseph Buisson Papers are available in the manuscript collec-

tions of the State Historical Society of Wisconsin. Buisson started
to organize some materials concerning the log- and lumber-rafting
days on the Mississippi River, and this small collection is the result
of his unfinished endeavor. Lists of boats in operation, the names of
captains and first pilots, plus some interesting descriptions of rafting
life are included.

A number of theses and dissertations deal with Wisconsin lumbering and related subjects. Most useful have been Duane Dale Fischer,
"The John S. Owen Enterprises," unpublished Ph.D. dissertation, University of Wisconsin, 1964; Dale Arthur Peterson, "Lumbering on the
Chippewa—The Eau Claire Area, 1845–1885," unpublished Ph.D. dissertation, University of Minnesota, 1970; and James Bruce Smith, "The
Movements for Diversified Industry in Eau Claire, Wisconsin, 1879–
1907: Boosterism and Urban Development Strategy in a Declining
Lumber Town," unpublished master's thesis, University of Wisconsin,
1967. Four additional studies provided occasional assistance: Duane
D. Fischer, "The Disposal of Federal Lands in the Eau Claire Land
District of Wisconsin, 1848–1925," unpublished master's thesis, University of Wisconsin, 1961; Willard F. Miller, "A History of Eau
Claire During the Civil War," unpublished master's thesis, University
of Wisconsin, 1954; Charles E. Twining, "Lumbering and the Chippewa River," unpublished master's thesis, University of Wisconsin,
1963; and Thomas J. Vaughan, "Life of the Wisconsin Lumberjack,
1850–1880," unpublished master's thesis, University of Wisconsin,
1951

BOOKS AND ARTICLES

A Century of Wisconsin Agriculture. Wisconsin Crop and Livestock
    Reporting Service, Bulletin No. 290. Madison: Published by the
    State, 1948.
Albion, Robert G. *Forest and Sea Power.* Cambridge: Harvard Uni-
    versity Press, 1926.
*American Lumbermen: The Personal History and Public and Busi-
    ness Achievements of One Hundred Eminent Lumbermen of the
    United States.* Chicago: American Lumberman, 1905.
Bailey, William F., editor. *History of Eau Claire County Wisconsin.*
    Chicago: G. F. Cooper & Company, 1914.
————. *Digest of Decisions of the Supreme Court of Wisconsin.* 3
    volumes. Chicago: Callaghan & Company, 1902.
Barland, Lois. *Sawdust City.* Stevens Point, Wisconsin: Worzalla
    Publishing Company, 1960.
Bartlett, William W. *History, Tradition and Adventure in the Chip-
    pewa Valley.* Chippewa Falls, Wisconsin: Chippewa Printery,
    1929.
Blair, Walter A. *A Raft Pilot's Log.* Cleveland: Arthur H. Clark
    Company, 1930.

Boyd, Robert K. "Up and Down the Chippewa River." *Wisconsin Magazine of History*, volume 14 (March, 1931).

Brown, Nelson Courtlandt. *Logging, the Principles and General Methods of Harvesting Timber in the United States and Canada.* New York: John Wiley & Sons, Inc., 1949.

Bryant, Ralph Clement. *Logging: The Principles and General Methods of Operation in the United States.* New York: John Wiley & Sons, Inc., 1913.

Cameron, Jenks. *The Development of Governmental Forest Control in the United States.* Baltimore: Johns Hopkins Press, 1928.

Carroll, Charles F. *The Timber Economy of Puritan New England.* Providence: Brown University Press, 1973.

Carstensen, Vernon. *Farms or Forests: Evolution of a State Land Policy for Northern Wisconsin, 1850–1932.* Madison: University of Wisconsin College of Agriculture, 1958.

Chippewa County Wisconsin. 2 volumes. Chicago: S. J. Clarke Publishing Company, 1913.

Chippewa Valley Business Directory for 1873. Eau Claire, Wisconsin: Free Press Print, 1873.

Clark, James I. *Cutover Problems: Colonization, Depression, Reforestation.* Madison: State Historical Society of Wisconsin, 1956.

————. *Farming the Cutover: The Settlement of Northern Wisconsin.* Madison: State Historical Society of Wisconsin, 1956.

Cochran, Thomas C., and Miller, William. *The Age of Enterprise.* New York: Macmillan Company, 1943.

Columbian Biographical Dictionary and Portrait Gallery of the Representative Men of the United States. Chicago: Lewis Publishing Company, 1895.

Conard, Howard L., editor. *Encyclopedia of the History of Missouri.* New York: Southern History Company, 1901.

Current, Richard N. *Pine Logs and Politics: A Life of Philetus Sawyer, 1861–1900.* Madison: State Historical Society of Wisconsin, 1950.

Curtis, John C. *The Vegetation of Wisconsin.* Madison: University of Wisconsin Press, 1959.

Defenbaugh, James Elliot. *The History of the Lumber Industry of America.* Chicago: American Lumberman, 1907.

Durant, Edward W. "Lumbering and Steamboating on the St. Croix River." *Minnesota Historical Collections*, volume 10, pt. 2. St. Paul: Published by the Society, 1905.

Ellis, Albert G. "Upper Wisconsin Country." *Wisconsin Historical Collections*, volume 3. Madison: Published by the Society, 1857.

Engberg, George B. "Lumber and Labor in the Lake States." *Minnesota History*, volume 36 (1959).

Everest, D. C. "Reappraisal of the Lumber Barons." *Wisconsin Magazine of History*, volume 36 (Autumn, 1952).

Fernow, Bernhard E. *History of Forestry.* Toronto: University Press, 1907.

————. "Difficulties in the Way of Rational Forest Management by Lumbermen." State of Wisconsin Horticultural Society, *Transactions, 1893.* Madison: Published by the State, 1893.

Fish, Carl Russell. "Phases of Economic History in Wisconsin, 1860–1870." State Historical Society of Wisconsin, *Proceedings, 1907.* Madison: Published by the Society, 1907.

Forbes, Reginald D. *Forestry Handbook.* New York: Ronald Press Company, 1955.

Forrester, George, editor. *Historical and Biographical Album of the Chippewa Valley.* Chicago: A. Warner, Publisher, 1891–1892.

Fries, Robert F. *Empire in Pine: The Story of Lumbering in Wisconsin, 1830–1900.* Madison: State Historical Society of Wisconsin, 1951.

————. "The Mississippi River Logging Company and the Struggle for the Free Navigation of Logs, 1865-1900." *Mississippi Valley Historical Review.* Volume 35, No. 3 (December, 1948).

Gates, Paul W. *The Wisconsin Pine Lands of Cornell University: A Study in Land Policy and Absentee Ownership.* Ithaca, New York: Cornell University Press, 1943.

————. "Weyerhaeuser and the Chippewa Logging Industry." *Augustana Library Publications,* no. 26 (The John H. Hauberg Historical Essays). Rock Island, Illinois: Augustana Book Concern, 1954.

Glasier, Gilson G., editor. *Autobiography of Roujet D. Marshall, Justice of the Supreme Court of the State of Wisconsin, 1895–1918.* 2 volumes. Madison: Democrat Printing Company, 1923.

Glover, Wilbur H. "Lumber Rafting on the Wisconsin River." *Wisconsin Magazine of History,* volume 25 (December, 1941–March, 1942).

Goodstein, Anita Shafer. *Biography of a Businessman: Henry W. Sage, 1814–1897.* Ithaca: Cornell University Press, 1962.

Greeley, William B. *Forests and Men.* New York: Doubleday, 1951.

Gregory, John C., editor. *West Central Wisconsin: A History.* 4 volumes. Indianapolis, Indiana: S. J. Clarke Company, Inc., 1933.

Gulick, Luther H. *American Forest Policy: A Study of Government Administration and Economic Control.* New York: Duell, Sloan & Pearce, 1951.

Hartsough, Mildred L. *From Canoe to Steel Barge on the Upper Mississippi.* Minneapolis: University of Minnesota Press, 1934.

Hauberg, John H. *Weyerhaeuser & Denkmann: Ninety-five years of Manufacturing and Distribution of Lumber.* Rock Island, Illinois: Augustana Book Concern, 1957.

Haugen, Nils P. *Pioneer and Political Reminiscences.* Madison: State Historical Society of Wisconsin, n.d. [Reprinted from the *Wisconsin Magazine of History,* volumes 11, 12, and 13.]

Helgeson, Arlan. *Farms in the Cutover: Agricultural Development in Northern Wisconsin.* Madison: State Historical Society of Wisconsin, 1962.

Herron, William H. "Profile Surveys of Rivers in Wisconsin." *United States Geological Survey, Water-Supply Paper 417.* Washington, D.C.: Government Printing Office, 1917.

Hidy, Ralph W.; Hill, Frank Ernest; and Nevins, Allan. *Timber and Men: The Weyerhaeuser Story.* New York: Macmillan Company, 1963.

Holt, William Arthur. *A Wisconsin Lumberman Looks Backward.* Oconto, Wisconsin: privately printed, 1948.

Horn, Stanley F. *This Fascinating Lumber Business.* New York: Bobbs-Merrill Company, 1951.

Hotchkiss, George W. *History of the Lumber and Forest Industry of the Northwest.* Chicago: G. W. Hotchkiss & Company, 1898.

Hurst, James Willard. *Law and Economic Growth: The Legal History of the Lumber Industry in Wisconsin, 1836–1915.* Cambridge: Belknap Press of Harvard University Press, 1964.

Ingram, Orrin Henry. *Autobiography.* Eau Claire, Wisconsin: privately printed, 1912.

Ise, John. *The United States Forest Policy.* New Haven: Yale University Press, 1920.

Jensen, Vernon H. *Lumber and Labor.* New York and Toronto: Farrar & Rinehart, Inc., 1945.

Kaysen, James P. *The Railroads of Wisconsin, 1827–1937.* Boston: The Railway and Locomotive Historical Society, Inc., 1937.

Keyes, Willard. "Journal of Life in Wisconsin One Hundred Years Ago." *Wisconsin Magazine of History,* volume 3 (March–June, 1920).

Kleven, Bernhardt J. "The Mississippi River Logging Company." *Minnesota History,* volume 27 (1946).

Knapp, Henry E. "The Trails from Lake Pepin to the Chippewa." *Wisconsin Magazine of History,* volume 4 (September, 1920).

Kohlmeyer, Frederick W. "Northern Pine Lumbermen: A Study in Origins and Migrations." *Journal of Economic History* volume 16 (December 1956).

———. *Timber Roots: The Laird, Norton Story, 1855–1905.* Winona, Minnesota: Winona County Historical Society, 1972.

Kuehnl George J. *The Wisconsin Business Corporation.* Madison: University of Wisconsin Press, 1959.

Lamb, Charles F. "Sawdust Campaign." *Wisconsin Magazine of History,* volume 22 (September, 1938).

Lapham, Increase A.; Knapp, Joseph G.; and Crocker, Hans. *Report on the Disastrous Effects of the Destruction of Forest Trees Now Going on So Rapidly in the State of Wisconsin.* Madison: Published by the State, 1867.

Larson, Agnes M. *History of the White Pine Industry in Minnesota.* Minneapolis: University of Minnesota Press, 1949.

Laurent, Francis W. *The Business of a Trial Court. 100 Years of Cases: A Census of the Actions and Special Proceedings in the*

Circuit Court for Chippewa County, Wisconsin, 1855–1954. Madison: University of Wisconsin Press, 1959.

Lillard, Richard G. *The Great Forest.* New York: Alfred A. Knopf, 1947.

Lincoln, Ceylon C. "Personal Experiences of a Wisconsin River Raftsman." State Historical Society of Wisconsin, *Proceedings, 1910.* Madison: Published by the Society, 1910.

Locklin, D. Phillip. *Economics of Transportation.* Chicago: Business Publications, Inc., 1935.

Lockwood, James H. "Early Times and Events in Wisconsin." *Wisconsin Historical Collections,* volume 2. Madison: Published by the Society, 1856.

Loehr, Rodney C. "Saving the Kerf: The Introduction of the Band Saw Mill." *Agricultural History,* volume 23 (July, 1949).

Lower, Arthur Reginald, *et al. The North American Assault on the Canadian Forest.* Toronto: Ryerson Press, 1938.

————. *Great Britain's Woodyard: British America and the Timber Trade, 1763–1867.* Montreal and London: McGill-Queen's University Press, 1973.

Lundsted, James E. "Log Marks, Forgotten Lore of the Logging Era." *Wisconsin Magazine of History,* volume 39 (Autumn, 1955).

Marshall, Roujet D. [See Glasier, Gilson G.]

Martin, Lawrence. *The Physical Geography of Wisconsin.* Madison: Published by the State, 1932.

Maybee, Rolland H. *Michigan's White Pine Era, 1840–1900.* Lansing: Michigan Historical Commission, 1960.

Merk, Frederick. *Economic History of Wisconsin During the Civil War Decade.* Madison: State Historical Society of Wisconsin, 1916.

Moran, Joe A. "When the Chippewa Forks Were Driving Streams." *Wisconsin Magazine of History,* volume 26 (1942).

Nelligan, John E. "The Life of a Lumberman." *Wisconsin Magazine of History,* volume 13 (September, 1928–March, 1930).

Nesbit, Robert C. *Wisconsin: A History.* Madison: University of Wisconsin Press, 1973.

Norton, Matthew G. *The Mississippi River Logging Company.* N.p.: Published by the Author, 1912.

Putnam, Henry C. "The Forests of Wisconsin." State of Wisconsin Horticultural Society, *Transactions, 1893.* Madison: Published by the State, 1893.

Randall, Thomas E. *History of Chippewa Valley.* Eau Claire, Wisconsin: Free Press Print, 1875.

Raney, William R. "Pine Lumbering in Wisconsin." *Wisconsin Magazine of History,* volume 19, (1935).

————. *Wisconsin: A Story of Progress.* New York: Prentice-Hall Inc., 1940.

Rector, William Gerald. "From Woods to Sawmill: Transportation

Problems in Logging." *Agricultural History*, volume 23 (October, 1949).

————. "Working with Lumber Industry Records." *Wisconsin Magazine of History*, volume 33 (June, 1950).

————. *Log Transportation in the Lake States Lumber Industry, 1840–1918*. Glendale, California: A. H. Clark Company, 1953.

Reynolds, A. R. *The Daniel Shaw Lumber Company: A Case Study of the Wisconsin Lumbering Frontier*. New York: New York University Press, 1957.

Rodgers, Andrew Danny III. *Bernhard Eduard Fernow: A Story of North American Forestry*. Princeton: Princeton University Press, 1951.

Roth, Filibert. *On the Forestry Conditions of Northern Wisconsin*. Madison: Published by the State, 1898.

Schafer, Joseph. *A History of Agriculture in Wisconsin*. Madison: State Historical Society of Wisconsin, 1922.

Schmid, A. A. "Water and the Law in Wisconsin." *Wisconsin Magazine of History*, volume 45 (Spring, 1962).

Sherman, Simon A. "Lumber Rafting on the Wisconsin River." State Historical Society of Wisconsin, *Proceedings, 1918*. Madison: Published by the Society, 1918.

Sieber, George W. "Sawlogs for a Clinton Sawmill." *Annals of Iowa*. 3rd series, volume 37 (Summer, 1964).

Smith, Alice E. *Millstone and Saw: The Origins of Neenah–Menasha*. Madison: State Historical Society of Wisconsin, 1966.

————. *The History of Wisconsin. Volume I: From Exploration to Statehood*. Madison: State Historical Society of Wisconsin, 1973.

Smith, Leonard S. *The Water Powers of Wisconsin*. Madison: Published by the State, 1908.

Smith, William Rudolph. *Observations on the Wisconsin Territory, Chiefly on That Part called the Wisconsin Land District*. Philadelphia: Carey & Hart, 1838.

Solberg, Erling D. *New Laws for New Forests*. Madison: University of Wisconsin Press, 1961.

Springer, John S. *Forest Life and Forest Trees*. New York: Harper & Brothers, Publishers, 1851.

Stanchfield, Daniel. "History of Pioneer Lumbering on the Upper Mississippi and Its Tributaries." *Minnesota Historical Society Collections*, volume 9 (1901).

Stephenson, Isaac. *Recollections of a Long Life, 1829–1915*. Chicago: privately printed, 1915.

Stoveken, Ruth. "The Pine Lumberjacks in Wisconsin." *Wisconsin Magazine of History*, volume 30 (March, 1947).

Strahl, Osborn. "A Chippewa Good Samaritan." *Wisconsin Magazine of History*, volume 7 (March, 1924).

Swisher, Jacob A. *Iowa, Land of Many Mills*. Iowa City: State Historical Society of Iowa, 1940.

Thomson, Betty Flanders. "The Woods Around Us." *American Heritage*, volume 9 (August, 1958).

Thwaites, Reuben Gold. *Wisconsin: The Americanization of a French Settlement*. New York: Houghton Mifflin Company, 1908.

Timber Construction Manual. Ottawa: Canadian Institute of Timber Construction, 1961.

Turner, J. M. "Rafting on the Mississippi." *Wisconsin Magazine of History*, volume 23 (December, 1939–March, 1940–June, 1940) and volume 24 (September, 1940).

Twining, Charles E. "Plunder and Progress: The Lumbering Industry in Perspective." *Wisconsin Magazine of History*, volume 47 (Winter, 1963–1964).

Ullman, Edward Louis. *American Commodity Flow*. Seattle: University of Washington Press, 1957.

Vinette, Bruno. "Early Lumbering on the Chippewa." *Wisconsin Magazine of History*, volume 9 (June, 1926).

Walsh, Margaret. *The Manufacturing Frontier: Pioneer Industry in Antebellum Wisconsin, 1830–1860*. Madison: State Historical Society of Wisconsin, 1972.

Warder, John A. "Forests and Forestry in Wisconsin." State of Wisconsin Horticultural Society, *Transactions, 1880–81*. Madison: Published by the State, 1881.

Warren, George Henry. *The Pioneer Woodsman as He is Related to Lumbering in the Northwest*. Minneapolis: Press of Hahn & Harmon Company, 1914.

Wells, O. V. "The Depression of 1873–79, Historical Aspects of Agricultural Adjustment." *Agricultural History*, volume 2 (July, 1937).

Weyerhaeuser, Frederick K. *Trees and Men*. New York: Newcomen Society in North America, 1951.

Wheeler, W. Reginald. *Pine Knots and Bark Peelers*. New York: Ganis and Harris, 1960.

Willson, Lillian M. *Forest Conservation in Colonial Times*. St. Paul: Minnesota Historical Society, 1948.

Wood, Richard G. *A History of Lumbering in Maine, 1820–1861*. Orono, Maine: University Press, 1951.

Wyllie, Irvin G. "Land and Learning." *Wisconsin Magazine of History*, volume 30 (December, 1946).

Wyman, Walker D. *The Lumberjack Frontier: The Life of a Logger in the Early Days on the Chippeway*. Lincoln, Nebraska: University of Nebraska Press, 1969.

NEWSPAPERS AND TRADE JOURNALS

For most of the period involved in this study, the *Eau Claire Free Press* and the *Chippewa* (Chippewa Falls) *Herald* have been the important newspaper sources. In the pre-Civil War period, there are a few issues of the *Eau Claire Telegraph* available at the State Historical Society of Wisconsin. Occasional reference has also been made to other local papers, such as the *Chippewa Union and Times* in the late 1860's, and the *Eau Claire Daily Telegram* since 1900. The *Milwaukee Sentinel* offers the special advantage of an index at the Milwaukee Public Library, although the index is complete only to the 1870's.

The most important trade journals for those involved in Lake States lumbering operations were the *Northwestern Lumberman*, published at Chicago, and the *Mississippi Valley Lumberman and Manufacturer*, published at St. Paul. Reference to the first of these journals included the years 1876 to 1889, and the second, 1876 to 1896. The *Mississippi Valley Lumberman and Manufacturer* is available at the Minnesota Historical Society Library in St. Paul, and the *Northwestern Lumberman* at the State Historical Society of Wisconsin. Other newspapers and trade journals—all from Wisconsin unless noted—include:

> *Alma Journal*, 1867–1869.
> *Alma Weekly Express*, 1869–1879.
> *Burlington* (Iowa) *Post*, 1930.
> *Dunn County News*, 1871.
> *Durand Weekly Times*, 1870.
> *Milwaukee Free Press*, 1918.
> *Rice Lake Chronotype*, 1967.
> *Winona* (Minnesota) *Republican Herald*, 1905–1907.
> *Wisconsin Lumberman*, 1874–1875.

PUBLIC DOCUMENTS

Most of the statutory materials and legal publications pertinent to lumbering subjects originate at the state and local level. In the course of this study, primary use was made of the State of Wisconsin *Private & Local Laws*, between the years of 1857 and 1872; Wisconsin *Laws*, from 1873 to 1883; Wisconsin *Statutes*, 1871; and Wisconsin *Revised Statutes*, 1867 and 1878. Some additional information was obtained from the official *Journal of the Senate* and *Journal of the Assembly* of the State of Wisconsin from 1867 to 1875.

Largely as a result of questions relating to the navigability of logging and rafting streams, federal statutes and federal courts assume some occasional importance. In this regard, a useful reference is the *Extracts from the Reports of Decisions of the Supreme Court of the United States concerning Navigable Waters, Riparian Proprietors, Bridges, etc.*, compiled in the Office of the Chief of Engineers, United States Army, Washington, D.C.: Government Printing Office, 1882. Because of the Corps of Engineers' long involvement with stream improvement projects, the House and Senate Executive Documents containing the Engineer's Reports often provide much detailed information about streams, their characteristics, and their usage.

Other documents consulted include:

United States Census Bureau. Ninth Census of the United States. *The Statistics of the Wealth and Industry of the United States.* Washington, D.C.: Government Printing Office, 1872.

————. Tenth Census of the United States. *Statistics of the Population of the United States.* Washington, D.C.: Government Printing Office, 1883.

————. Twelfth Census of the United States. *The Statistics of the Wealth and Industry of the United States.* Washington, D.C.: Government Printing Office, 1902.

United States Department of Agriculture. Weather Bureau Bulletin. *Summaries of Climatological Data by Sections.* Volume 2. Washington, D.C.: Weather Bureau, 1912.

A Synoptical Index of the General and Local Laws of Wisconsin From the Organization of the Territory to 1873 Inclusive. Madison: Published by the State, 1873.

INDEX

CHARLES E. TWINING, a native of Kansas, holds a doctorate from the University of Wisconsin. He teaches history at Northland College in Ashland, Wisconsin.